International Poultry L

PRACTICAL
POULTRY KEEPING

Battery cages

Note the birds are four per cage, little space, no perches, beaks trimmed to avoid canibalism and can only eat, drink and lay. The legal position is that cages will be banned in about 10 years so alternative, economic systems have now to be developed. Standard breeds could now be brought back to free range and semi-intensive systems of management.

PRACTICAL POULTRY KEEPING

11th Edition

Edited and Revised by

Dr Joseph Batty

Chairman: World Bantam & Poultry Society

Northbrook Publishing Ltd

Beech Publishing House

Station Yard

Elsted Marsh

Midhurst

West Sussex GU29 OJT

ISBN 1-85736-384-1

First published 1943

for *Poultry World.*

Revised and Expanded.

New Edition (11th) 2,000

British Library Cataloguing-in-Publication Data

A catalogue record for this book is available from the British Library.

Northbrook Publishing Ltd

Beech Publishing House

Station Yard

Elsted Marsh

Midhurst

West Sussex GU29 OJT

Contents

METRICATION AID TABLE

IMPERIAL MEASUREMENT = METRIC MEASUREMENT

Length	1 inch	=	2.54 cm
	1 foot	=	30.48 cm
		or	0.304 m
	1 yard	=	0.9144 m
Area	1 square foot	=	0.092903 m^2
	1 acre	=	0.404686 ha
Volume	1 cubic foot	=	0.0283168 m^3
	1 pint	=	0.568245 litre
	1 gallon	=	4.54596 litre
Weight	1 ounce	=	0.028349 kg
		or	28.349 g
	1 pound	=	454 g
	1 hundredweight	=	50.8023 kg
	56-lb	=	25.4 kg
Miscellaneous	1 joule	=	0.239 cal
	1 kilo calorie/pound	=	0.009223 Mj/kg
	1 mile	=	1.609344 km
Terms	centimetre	=	cm
	metre	=	m
	square metre	=	m^2
	hectare	=	ha
	cubic metre	=	m^3
	gram	=	g
	kilogram	=	kg

Temperature conversion

Centigrade to Fahrenheit
$$^\circ C + 40 \times {}^9\!/_5 - 40 = {}^\circ F$$

Fahrenheit to Centigrade
$$^\circ F + 40 \times {}^5\!/_9 - 40 = {}^\circ C$$

(vi)

INTRODUCTION

During the past 40 years or so, drastic changes have been seen in poultry farming. No longer are hens mere inhabitants of the farmyard, existing only to provide pin money for the farmer's wife; now they are the focal point of a vast array of technical research and knowledge that enables them to be managed on scientific lines as a sound business enterprise. Today, the hen is the product of a complex breeding programme worthy of the best feeding and management that can be given. Unfortunately, it can be argued that the process of breeding hybrids and raising them indoors, as well as keeping them in laying cages for the rest of their short lives, has gone too far. There is now general admission that the birds kept and raised intensively suffer from many man-inflicted ailments such as bronchitis, foot injuries from wire cage floors, brittle bones. breast blisters, and stress and strain from any faults in the controlled environment. There are also problems from birds pecking each other, leading to cannibalism, and whilst beak trimmimg or spectacles can reduce this problem, this is an undesirable practice.

In Europe the so-called hybrids in cages are now in the phase of changing over to more acceptable systems, such as free range, barn or aviary systems. Unfortunately, the hybrids were never meant for free range, so with their specially developed small bodies, intended to eat the smallest amount of food for maximum egg production, they are not really suitable for outdoor living and production. Now therefore is the time to turn back to the standard breeds such as Leghorns, Minorcas, Rhode Island Reds, Plymouth Rocks, Barnevelders, Sussex, Wyandottes and many others. They are still available and have the anti-bodies for protection for foraging and producing on free range. The strains that still lay well have to be located and brought back as quickly as possible.

The more specialized markets also have to be developed such as ducks, geese, turkeys, guinea fowl, and ostriches, thus giving the poulty farmer a wider market. Even standard bred poultry and bantams offer possible areas of development .But in all these possible areas of production and marketing the welfare aspects must be known and applied.

The poultry industry is now one of the main agricultural interests, with a turnover second only to the dairy industry. Despite the expansion that has already been achieved, however, the consumption of poultry meat and eggs in this country is still low compared with some other countries, so that the limits of expansion have by no means been reached. In fact, there are still immense possibilities, provided the industry continues to develop on the right lines.

Practical Poultry Keeping

In the first edition of this book it was pointed out that there were two kinds of poultry keeping: **Exhibition,** where birds are bred for form and beauty, **and Utility,** known today as the commercial poultry industry, where economic egg production and table poultry are the main aims. To some these sections are separated by objects and ideals that are irreconcilable. No doubt there will always be the diehards on both sides who will maintain this attitude, but the common sense view seems to be that each side has something to give the other, especially now battery cages are to be phased out in the next 10 years.

This book attempts to set out the general principles of practical poultry keeping. Some of the ideas are not entirely British but have been imported from the United States and other countries. There are many other new ideas which have been introduced since the book was last revised. These will be of interest to some and of economic importance to those engaged in poultry keeping. It is hoped they may be a stimulus to achieve greater efficiency from the production of meat and eggs.

Today, the geneticists and mathematicians have most of the answers to complex breeding problems. Genes governing the inheritance of commonly important economic characteristics have been studied. New breeding and selection techniques have led to the introduction of the 'hybrid' chicken, not only for egg production but meat production also. This has resulted in cheap eggs and meat, but to the detriment of the birds. The scientists have tried to move to far from the natural methods so they are open to the accusation that factory farming is being used to exploit the different types of poultry.

Correct nutrition of the modern chicken quite definitely influences economic characteristics more than any other external factor. It is also the most important and expensive item in the costs of producing eggs and poultry meat, and for this reason the poultry farmer must ensure that the most efficient use is made of the food. Knowledge of the fowl's nutrient requirements has increased tremendously during the past half century with the result that 1 kg of liveweight can be produced economically on 2 kg of food and 12 eggs on 1.8 kg of food. The information in this book is intended to give poultry keepers a brief insight into the fowl's requirements and, more importantly, practical feeding advice. Although the saving through using modern systems of intensive management is considerable compared with other systems, such as free range housing, the welfare requirements mean that these other systems must be adopted once more and made as efficient as possible.

RE-INTRODUCING THE STANDARD BREEDS

The laying tests conducted when standard bred poultry were still kept, showed that many breeds could produce 250 eggs as a norm; many produced more. With trap nesting the best layers could be selected.

There is a further advantage, over and above the welfare aspects, because poultry farmers who are flexible can meet the requirements of a local market or produce rather special eggs or meat at a gourmet level. For example, consumers prefer the very dark brown eggs (coffee coloured) laid by Barnevelders, Croad Langshans, Welsummers and Marans, and the poultry farmer can cater for this market on free range and command premium prices. On a larger scale a farmer could produce all white eggs, preferred for commercial purposes, thus using the splendid layers from the Mediterranean breeds such as Leghorns, Anconas or Minorcas.

Even exhibition birds, including bantams, have great potential. The author knows one breeder who produced 500 chicks for sale each year and these are sold at a handsome profit.

Examples of Breeds

The possible breeds to keep come to more than 50 and bantams are popular in just as many breeds. The Mediterranean, light breeds lay white eggs and are usually suitable for commercial production. The Heavy breeds lay a colour which is described as 'tinted' which means light to fairly dark brown. A few breeds, mentioned above, lay deep brown eggs.

Heavy

Australorp
Barnevelder
Brahmas
Cochin
Croad Langshan
Dorking
Faverolles
Houdans
Marans
Orpington
Plymouth Rock
Rhode Island Red
Sussex
Wyandotte

Light/Medium

Ancona
Andalusian
Hamburgh
Leghorn
Minorca
Old English Game
Poland
Scots Grey
Silkie
Welsummer

There are other breeds, but these are not likely to be commercially viable. Some are now quite rare.

Chapter 1

BREEDING

INTRODUCTION

Reproduction of fowls requires the mating of the two sexes so that male and female reproduction cells, i.e. sperm and ovum, may unite to give rise to a new individual. In the male bird the spermatozoa are produced by two testes; in the female ova are produced by the ovary. Another important function of these glands is to secrete substances known as hormones or chemical messengers. Their presence is necessary for the normal development and well-being of the various body structures.

From Fig. 1.1 it can be seen that the male reproductive system consists of two testes. Each testis is composed of tubules in which the spermatozoa are produced. The sperm leave the testes via the small epididymus (tube) and from there pass to the cloaca via the vas deferens. At mating the semen containing the sperm passes through papillae to the copulatory organ. Each sperm consists of a head in which is housed the nucleus, a 'middle piece' and a long tail.

The female reproductive system (Fig. 1.2) consists of an ovary and oviduct; the ovary produces the ova, and the oviduct transports them to the cloaca. It is during transportation through the oviduct that the different layers of albumen or egg white, shell membranes and shell are laid down around the ovum. The chicken embryo which develops on the surface of the yolk obtains its nourishment from the different parts of the egg. The shell provides protection and being porous, allows exchange of oxygen and carbon dioxide.

The chicken has only one ovary situated on the left-hand side, and

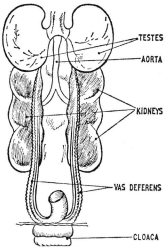

Fig. 1.1 Male reproductive system

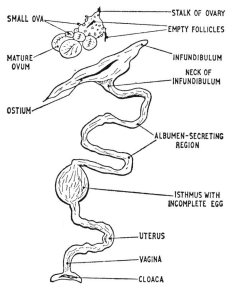

Fig. 1.2 Female reproductive system

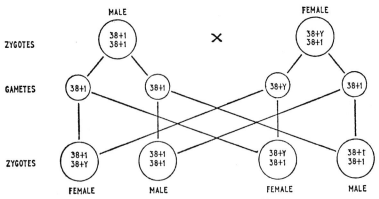

Fig. 1.3. Factors determining sex of offspring

there is therefore only one oviduct leading from the ovary to the cloaca. There are many thousands of ova on and in the laying bird's ovary, but they do not all develop, many of them degenerating before being observed by the naked eye.

Fertilisation

After mating has occurred, the spermatozoa travel up the oviduct using their long tails to propel them. Fertilisation occurs when a spermatazoon penetrates the germinal disc of the ripe ovum. The site of fertilisation is thought to be at the top of the oviduct in the infundibulum. Approximately 21 days after fertilisation the embryo develops completely and hatches.

Inheritance

The reproduction cells of the male and female chicken are known as *gametes*. Thus, a sperm is a gamete as also is an ovum. When mating occurs these reproduction cells fuse or join to form a *zygote*.

Each gamete contains *chromosomes*. It is these which are transmitted from the parent to the offspring. Chromosomes do not contain blood, so blood is not passed from parent to offspring. The number of chromosomes in the fowl is constant and as they occur in pairs are usually expressed this way; in chickens the number is thought to be 39 pairs. Chromosomes are of two distinct types: sex chromosomes and autosomes. The number of autosomes remains constant irrespective of

6

Two of the most popular utility breeds, the Light Sussex (above) and the Rhode Island Red (below). Both are heavy breeds and are very useful for crossing for either table poultry or egg production

sex, but sex chromosomes differ in respect of sex. Thus, the male chicken has two sex chromosomes and is called *homogametic,* the female only one and called *heterogametic.* In place of the other female sex chromosome is what is known as a *Y-chromosome.* In poultry it is the female which decides the sex of the offspring. Fig. 1.3 will make this clear.

It is important to remember that the chromosomes are the bearers of the determiners of the hereditary characteristics. As there are many hundreds of characteristics and only a comparatively few chromosomes it is obvious that each chromosome must be responsible for the development of many characteristics. Each chromosome contains *genes,* and it is these which control the development of characteristics such as egg production, plumage colour, etc.

Inheritance is therefore transmitted from parent to offspring; the characteristics are not bodily transmitted. It is the ability or power to develop these characteristics which is inherited, e.g. the ability to lay a large number of eggs.

It would be as well for the reader to realise that with economic characteristics the environment can play a very great role in masking or potentiating the effect of inheritance. These factors are known as relationships between heredity and environment. This explains the variations which exist in a genetically uniform flock of laying birds.

Breeding for economic characteristics

The main economic function in this country is to produce eggs and meat. The efficiency with which these utility operations can be carried out depends mainly upon genetic factors but is influenced to a considerable degree by the environment: for example, nutrition, housing and management. Inheritance of economic characteristics is dependent upon many gene pairs acting in various ways.

EGG PRODUCTION

There are six main characteristics involved in egg production:
- (a) Fecundity (number of eggs)
- (b) Sexual maturity
- (c) Intensity
- (d) Broodiness
- (e) Persistency
- (f) Feed efficiency

Related to these factors in many ways are the characteristics of egg size and shell quality.

Characteristics involved in breeding are fertility, hatchability and viability.

Breeding for egg production thus involves all the factors mentioned. They will now be considered in the order mentioned.

Fecundity

This means the number of eggs produced in a certain period of time, regardless of other characteristics. The pullet year, or 500 days from date of hatch, is usually used as the criterion of measurement. The most satisfactory method for determining egg numbers is that based on hen-housed production, for any flock with a good record of production must have been good in other respects: for example, low mortality.

Sexual maturity

Early and late sexual maturity have long been recognised as heritable characteristics, regardless of time of hatching, feeding and environment. Selection for early sexual maturity with precocity is relatively easy and can be done by handling individual birds. It does not need trapnesting to determine or measure it. When a period of 500 days from date of hatch is used as a measurement criterion then late maturing birds will generally give poorer results than those maturing early.

Intensity

Intensity of production is usually determined over a short period and was recently measured over a winter period of, say, 6 months. Whatever time of the year is used it is important to use trapnest facilities. A standard of not less than 26 eggs per month should be used over a minimum period of 3 months in winter and 4 months in the spring.

Broodiness

Modern light breeds and crosses are invariably less broody than the heavier Asiatic breeds and crosses. Broodiness can occur in the lighter Mediterranean breeds, depending upon strain and not breed. There is no justification to support the theory that an occasional rest from broodiness does a bird good. The genetic nature of this characteristic is such that broodiness, although apparently eliminated, may make an appearance through no obvious fault of the breeder.

Intensive breeding unit showing equipment layout

Small pedigree breeding pen to produce the female line

Persistency

This term refers to the length of the first laying year. High persistency is a most valuable trait which is easy to measure and does not necessarily require the use of trapnests. Its economic importance is obvious, especially when the higher egg prices occur at the end of the laying year or the bird has the ability to lay for 14 or 15 months.

Egg size

This is a most important factor in determining profitable commercial egg production. Egg size is therefore of considerable concern to most breeders; it usually increases rapidly up to 10-12 months of age, at which time physical maturity is reached. It is known that all small birds tend to lay small eggs, but not by any means do all large birds lay large eggs.

Shell quality

To obtain good shell quality the thickness and strength must be good. Not only does poor shell quality effect hatchability, thin-shelled eggs being difficult to hatch, but it also adversely affects the commercial egg producers' packing station returns. Inheritance plays quite a large part in determining shell thickness. It is not of course entirely a matter of breeding, for nutrition is also involved; there must be a proper balance of calcium, phosphorus, manganese and vitamin D3 in the birds' diet.

Fertility

It does not appear that heredity has much effect on fertility. The most important points to consider are management factors such as housing and nutrition.

Hatchability

The difference between good and bad hatches is primarily governed by heredity, although feeding and incubation obviously play major roles. Hatchability is defined as the percentage hatch of all fertile eggs. More recently hatchability has been calculated on the percentage of all eggs set. This is because it is claimed that all eggs produced from a healthy breeding flock are in fact fertile, and eggs commonly called infertile are not but are caused by early embryonic mortality.

Viability

In relation to certain diseases of major importance genes for resistance and probably immunity do exist, and in the continued absence of satisfactory control by means of drugs or management the ultimate solution to the problem will depend on the breeding of resistant strains of poultry. The problem is not a simple one, for inherited resistance to one disease does not always mean a conferred resistance to another disease. With leucosis it is known that certain strains have greater resistance than others to the disease. The difficulty arises when the flock has to be exposed to a predetermined level of infection in order to measure resistance or susceptibility. Few small breeders can afford such expensive programming.

HYBRIDS

Hybrids are produced by crossing two strains of parents which, though they differ genetically, complement each other in respect of gene make-up. Crossmating is stated, but whilst this is true, it does not necessarily infer crossing two different breeds or even varieties. It is essentially a matter of mating to exploit complementation and may take place between strain crosses within a breed or within a variety. It is important to realise that not every crossbred pullet possesses hybrid-vigour. The term *hybrid* does not necessarily apply only to laying birds but embraces table poultry, including ducks and turkeys. The use of the word in poultry breeding does not necessarily indicate a superiority over birds not bearing this prefix. Indeed, it can be said that many small breeders with closed flocks are producing a product of considerable merit although not called hybrid. There should always be a demand for this type of high-quality purebred stock as a basis for producing new hybrids.

There are currently three recognised systems used for producing hybrids:

(a) Selected strain crossing
(b) Incrossing and incrossbreeding
(c) Reciprocal recurrent selection

These will now be described.

Selected strain crossing

This method is based on parent strains that have been selected specifically for their ability to 'nick' (complement each other) in certain combinations. Usually, a number of different strains are tested, not necessarily all of the same breed. If eight were used they could be mated in 8 x 7 = 56 different combinations, including reciprocals. Breeders working on a small scale would ignore the reciprocals and use 28 combinations, not 56.

The system is simple to operate, as only samples of progeny produced need be used, say 100. They are tested under normal conditions and recorded. With good repeatability the reciprocal is tested. If this is successful then both strain crosses can be marketed as hybrids.

Incrossing and incrossbreeding

Also known as the 'Wallace' breeding system, this method is said to achieve three things:

(a) It discards useless material.
(b) It promotes breeding uniformity and repeatability of performance.
(c) It leads to degeneration in the basic material.

The system, as the name implies, involves close breeding. The larger the breeder's resources the more rapid the process can afford to be. With restricted resources half-brother-half-sister matings are used. Many of the inbred lines die out, others prove so poor that they have to be discarded. The surviving lines are crossed in as many combinations as possible to produce two-way crosses. These are tested under commercial conditions and against good competitors' stock.

Obviously, this system is expensive to practise and is therefore out of the question for small breeders.

Reciprocal recurrent selection

This is a more modern method of producing hybrid chickens. The system involves the use of two different stocks either of the same or different breeds. After selection of two flocks they must remain closed to outside stock or 'new blood'.

First, selected cocks of both flocks are progeny tested against the opposite flock of selected pullets, selection being based on individual performance. From the results of these matings it is possible to rank every crossmating in both directions according to individual sires and dams in each stock. In the following year the most proven sires are mated to the proven females to reproduce them as purebreds. The first 2 years' work is then repeated. Thus the system falls into 2-year cycles:

progeny testing one year and pure stock reproduction in the following year.

The main advantage of reciprocal recurrent selection is its absence of the need for inbred lines. The ideal flock size appears to be 1,000 pullets, based on 8-12 sires. This breeding system provides the smaller breeder with an answer to the larger breeding organisations in this country.

Hybrid table chicken

The chicken-broiler industry, like the egg reproduction side, is involved in the production of hybrids. The standards or requirements of each are, of course, poles apart, although the system used to produce the end product may be similar.

The characteristics which the breeder of broiler parent stock has to consider may number as many as 15. A few are: white shanks or flesh, high meat/bone ratio, good breast width, succulent flesh, quick feathering, rapid growth, good liveability, good egg production, fertility and hatchability. The importance of 'nicking' for egg production, hatchability and liveability are just as important in breeding for broilers as for egg production, as each plays a role in determining production costs and overall profit of the enterprise. It cannot be overstressed that with the economic characteristics mentioned there is a need for a constant environment and sound management when new strains are being tested and selected.

Assessing the economic performance in comparison with the conventionally bred chicken is not simple. By definition, the modern hybrid should be superior, but the reader is advised to make reference to the random sample tests conducted in this country and abroad. The results of these tests make useful guides in choosing between one commercial-laying bird and another.

To clarify a number of terms used in the section dealing with hybrids the reader may find the following of interest.

Progeny testing

Characteristics such as egg production and viability have what are called a *low hereditability*. This means that for steady progress to be made a selection of stock must be based on the records of families and not of one individual. To do this the breeder makes use of progeny testing. The principles are simple. To gain knowledge of the value of

14

one family or line the breeder would not select one bird and base his selection programme on this one result. He selects as many birds from the 'line' as possible, knowing that each bird's performance adds more to the overall picture of the line. By increasing the number of offspring the breeder increases the number of independent samples of the parents' genes and so obtains a progressively clearer picture of the genetic make-up of those parents. In progeny testing no culling is allowed. Should excess birds be available then they must be eliminated by a method of random sampling. The progeny have to be trapnested to obtain the full information of the parents' ability to transmit the desired characteristics. Progeny testing is not cheap to carry out, for detailed recording is absolutely necessary.

When a bird has passed the progeny test no effort should be spared to use the bird as a breeder, distributing its characteristics as quickly and efficiently as possible.

Inbreeding

There are two systems of non-random mating: namely, inbreeding and outbreeding.

The main aim of inbreeding is to increase the probability that the young of future generations will be alike genetically in respect of the characteristics which are required in their progeny. The result of inbreeding is increased homozygosity (pureness), which is brought about by multiplication of what are called *recessive* and *dominant* genes. As most undesirable characteristics are recessive a lowering of stock quality may result unless the offending stock is very quickly discarded. Undesirable recessives with less noticeable effects may, of course, easily be overlooked. Economic characteristics which rely on many gene pairs for their expressions depend on inbreeding to establish the traits. No other mating system is capable of doing this. Inbreeding is the surest and quickest way of exposing what is in any given closed flock. If too close inbreeding is practised it can wipe out complete flocks. Therefore, the breeder wishing to expose his flock to this system of mating should, beginning with good stock, rely on half-brother-half-sister relationships. No other special skill is required except, of course, in maintaining efficient and accurate records of performance. Different stock respond differently to the effects of inbreeding. This is because the genetic make-up of the original material is different. There will be considerable variation to the response, some stocks not being able to tolerate it for

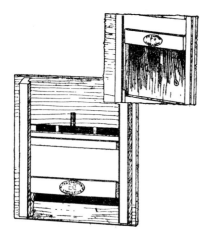

Fig. 1.4. A popular type of trapnest front, the top sketch shows the nest open for the entry of the bird and the lower with the front in a sprung position. The hanging wires seen in the top sketch form the mechanism of the nest; when a bird enters it pushes the wires inwards and releases the sliding plywood front which is slightly weighted at its base to ensure a free fall. Some types of front are hinged and buttoned for the purpose or providing easy removal of the bird. In hot weather many breeders fit wire-netted floors to the nests in place of the usual solid bottom

more than a few generations. Others may peter out at a later date through inability to reproduce themselves. (For further information on inbreeding and measurements on inbreeding the reader should consult the *M.A.F.F. on Poultry Breeding.*)

Outbreeding

Outbreeding is the opposite of inbreeding and refers to systems of mating in which the parent stock have a less-than-average relationship of each other. Outbreeding causes the population to become temporarily more uniform in outward performance, although due to loss of homozygosity (pureness) performance is lowered for a time.

The main use of outbreeding is to 'upgrade' stock by introducing fresh 'blood'. The reverse generally occurs. Because of this any new blood introduced to the flock should first be tested on a small sample of the main flock to determine its effect.

Trapnesting

The use of the trapnest is essential when progeny testing is being carried out. The date of starting the traps varies from breeder to breeder. Some commence trapping directly the first egg is laid; others start at a pre-determined date, depending on age of calculated genetic sexual maturity. The usual period is a full 48 weeks or 500 days from date of hatch. It is not essential to trapnest every day of the week as

The Rhode Island Red x Light Sussex mating (below) is one of the most popular sex-linked crosses. Chicks (above) are distinguished at day-old by down colour, the cockerels being white or silvery and the pullets buff or brown

past records of, say 5/7 give a fair degree of accuracy. Trapping less than 5 in 7 days is not recommended.

There are many types of trapnests. The most usual are fronts or doors that can be fitted to the nest boxes, the bird closing the shutter as it enters and yet not so large as to allow two birds to enter at the same time. The catch which holds the trapnest in position should be easily removed. Fig. 1.4 illustrates a typical trapnest front.

Lethal and sub-lethal genes

A lethal gene is one which causes the death of the developing embryo, whereas a sub-lethal gene generally causes deformity or death after the chick is exposed to the environment. Most lethal genes in poultry are recessive, inferring that their existence is made known only by studying the records of hatchability. Many lethal and sub-lethal genes exist in poultry, and it is important for the breeder to appreciate their existence. The presence of lethal genes is discovered mainly through inbreeding, and in normal circumstances the breeder would not recognise the deleterious effects of a recessive lethal in the pure state unless, as mentioned earlier, he studied carefully the records of hatchability. To eliminate a lethal carrier from the flock the breeder has to dispense with not only the 'carrier' parent, but also the progeny because many of them would carry the lethal in the impure state.

Collectively, all lethal and sub-lethal genes constitute an economic challenge to the poultry industry, the seriousness of which is probably not realised as broadly as it should be.

In 1949, 21 lethals were mentioned in chicken, 3 in turkeys and 1 in ducks. Of the 21 in fowls, 5 were sub-lethal.

Sex linkage

It has long been known that with certain matings it is possible to distinguish the sex of chicks at the time of hatching by the differences in down colour. For example, the cross between a Rhode Island Red cockerel and a Light Sussex hen gives rise to brown pullets and silver cockerels at hatching. This is commonly known as the 'gold cross silver' sex linkage. It is the silver gene of the female which is dominant, for as we have seen earlier in the chapter it is the female which decides the sex of the chick and carries only one sex chromosome. The reverse cross, namely Light Sussex cockerel with a Rhode Island Red hen gives rise, as we should expect, to all silver chicks. This criss-cross inheritance is not confined only to gold and silver feathered birds of the breeds named. It can be produced by crosses involving the Light Sussex with Buff Rocks, Brown Leghorns, Buff Orpingtons and New Hampshire Reds.

Sex linkage is not confined to only gold and silver plumaged birds but also applies to slow and fast feathering, barring and non-barring, and dark shanks and light shanks. Slow feathering is dominant to fast feathering, barring to non-barring and dark shanks to light shanks.

The practical importance of sex-linkage is well known to poultry men today and, in theory, quite a number of characteristics could be employed to help sex chicks, but, in practice, only a few are really suitable.

Physical selection of breeding stock

While the modern selection of breeding stock both for egg production and meat production demands the use of complex breeding systems and statistical analysis it must not be forgotten that physical handling of such birds still plays a part in the overall programme. What handling can do is to eliminate birds from the breeding flock which are physically incapable (no matter how good their records may appear) of making suitable breeding material. This section particularly applies to poultry keepers who are involved in what is termed *stock multiplication*. This is usually practised by the larger chick-producing organisations, which distribute the parent stock.

Selection is based on the physical handling qualities of the breeding stock at regular intervals during rearing and at point of lay. The success of the stock multiplier's work depends on his having vigorous and healthy stock. He must be prepared to kill off or otherwise dispose of every breeder that shows loss of vigour. Should he fail to recognise this he may fail to get the best possible return from his flock. The eye has been described as a mirror of vitality. This is true, for the eye reveals the bird's state of health. It should be bold and prominent with an alert appearance. Structural eye defects should be regarded with suspicion. The bird's stance should be good. The body should handle well for the bird or strain type. In many cases bodyweight ranges for the bird are given, and within reason they must be adhered to. Grossly underweight or thin birds should be discarded, for they may be suffering from disease.

The cockerels too should be fit and full of vigour. The birds that by their appearance, behaviour and handling qualities have every evidence of possessing an abundance of vigour should give a good account of themselves, at least in the sense that mortality should not be high, but of course there can be no assurance on this point. Occasionally, flocks that have passed a handling and observation test by poultrymen of long experience and ability, exhibit heavy mortality in the course of the breeding season. The so-called 'three-finger-exercises' practised by poultrymen of the old school have long been abandoned. Abandoned, not because they are worthless, but because they give no indication of the bird's ability to transmit economic characteristics.

The most convincing evidence of vigour is the ability of the bird to maintain a satisfactory level of egg production, to maintain or increase bodyweight during that time and to produce healthy progeny. Loss of weight may indicate that a bird is unable to assimilate sufficient food.

Age for mating

Precocious cockerels, and for that matter pullets, should not be used for breeding purposes. Both sexes, maturing at an early age, usually lack body size and may break down in the course of the breeding season. Cockerels should be fully matured before being used for breeding purposes. They should be 5-6 months old, the younger age for egg-producing strains and the latter for broiler breeding stock. Pullets may be one month younger in both examples.

Number of cockerels

The number of hens needed to mate with each cockerel will depend upon strain and the conditions under which the birds are kept. A vigorous light-type cockerel is able to service up to 15 hens, whilst with broiler breeders the number should be reduced to 10.

Chapter 2

INCUBATION

INTRODUCTION

The incubation of poultry eggs, either naturally (with broody hens) or artificially (in an incubator), cannot be performed or supervised efficiently without knowledge of the fundamental principles involved. To understand the physical requirements of incubation it is necessary first to have some knowledge of the structure of a hen's egg.

Structure of the egg

The egg is composed of five different parts (see Fig. 2.1), the percentages of which according to gross weight are given below:

Shell and shell membranes	12%
Albumen and chalazae	56%
Yolk	32%
	100%

Both the albumen and yolk differ widely in chemical composition. The percentage distribution (fresh weight) is given below:

	Whole egg	Albumen	Yolk
Water	73.7	87.77	49.0
Protein	13.4	10.00	16.7
Fat	10.5	0.05	31.6
Ash	1.0	0.82	1.5

Shell and shell membranes

The shell consists mainly of calcium carbonate and forms a protective layer to the egg's fluid content. It is porous to allow interchange of moisture, oxygen and carbon dioxide with the outside atmosphere. Because it is porous any dirty material on its surface may gain entry to the shell membranes. Therefore, only clean, uncontaminated eggs should be incubated.

There are two shell membranes, both of which are thin and tough

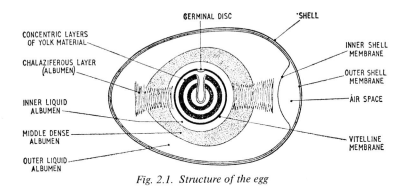

Fig. 2.1. Structure of the egg

Separating these membranes at the egg's broad end is the air space. It is from this air that the embryo, just before hatching, draws its first supply of oxygen.

Albumen and chalazae

The albumen, or white of the egg, is in four parts; three of these are in distinct layers around the yolk. The chalazae, on the other hand, attaches itself to the yolk surface and the middle layer of albumen. Its purpose can be compared with that of shock absorbers. It also maintains the yolk in a central position. The strands of chalazae are wound in opposite directions.

The yolk

The yolk, which originates from the ovary, is enclosed in a membrane known as the vitelline membrane. On the surface of the yolk is situated the female germ cell or blastodisc. Whilst the blastodisc is present in all eggs it only becomes active when fertilised by a male

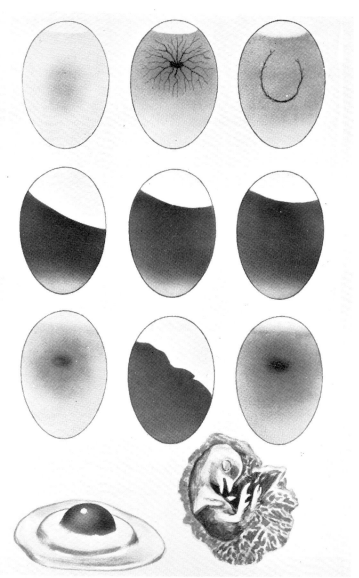

The egg during incubation: Top row (7th day) left, clear egg; centre, fertile egg; right, broken yolk. Middle row (14th day) left, egg dried too much – too little moisture; centre, correct drying; right, too little drying – excessive moisture. Bottom row left, dead embryo about 5th day; centre, ready to hatch 20th day, right, embryo died on about 14th day. Left, at bottom, the egg yolk and small white germinal spot; right, an embryo chicken at 14-15th day of incubation.

spermatozoon. The yolk itself serves as a rich store of food materials for the newly hatched chick. It is composed of concentric layers of dark-and light-coloured yolk.

DEVELOPMENT OF THE CHICK (EMBRYOLOGY)

After fertilisation in the hen's oviduct the single fertilised cell develops rapidly by a process of segmentation. A single cell forms two, the two four, the four sixteen, and so on. Development has thus commenced before the egg is laid. The term cell is in three layers, i.e. top, middle and bottom. It is these three layers which multiply to form the mature embryo.

By the fourth day of incubation the leg and wing buds are discernible; brain and nervous tissue is also seen. By the sixth day the main subdivisions of the legs and wings are evident. Feather tracts can be seen by the eighth day and by the ninth the embryo begins to look like a chick. Calcification of the bones is complete by the 15th day and the colour of the down is seen by the 13th day. On the 16th day all horny structures, i.e. beak, claws and leg scales, are well formed. It is on the 19th day that the chick commences to draw the yolk sac into its body cavity. Final development involves the muscle maturation in readiness for the strain of breaking out of the shell. When the shell is 'pipped 'the embryo's struggles cause it to rotate within the shell until the top of the shell is chipped off, enabling the embryo to break free.

Malpositions of embryos

Many embryos fail to hatch because they are in an incorrect position at the time of pipping the shell. The correct position is: body along the axis of the egg with head towards the broad end, facing right and under the right wing. In this position it will be facing the air space.

Malpositions encountered are: head between thighs, head in small end of egg, head to the left, body rotated, feet over head, head over wing and embryo across the egg. Careless handling of hatching eggs during the last week of incubation will have serious repercussions on the embryos. The critical periods in the embryo's development are between the third and fifth days and between the 18th and 19th days. A further critical period is between the 12th and 14th days of incubation, although this is less so than the other two periods.

Selection and handling of hatching eggs

To obtain best hatchability results, attention should be given to several points in the selection and handling. On the average, 18%-20% of all eggs incubated fail to hatch; thus any improvement on these figures must be an economic one.

The best hatching results are obtained with eggs weighing between 50-60 g. and since egg size is inherited selection on this factor will help maintain good egg size in the progeny. Shape of the eggs should also be given priority in selection, for this character is also inherited.

Misshapen eggs are also likely to have abnormal internal contents. The porosity of the shell determines the rate of moisture loss during holding and incubation. Since this characteristic is also inherited only eggs of good shell texture should be incubated. Naturally, eggs with hair cracks

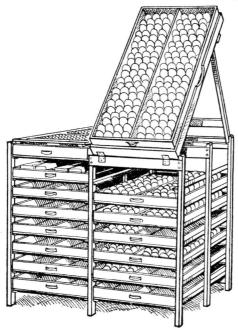

Fig. 2.2 The setting of incubator trays is facilitated by the use of a rack, which stands near the machine or could be adapted to run on small wheels. After the trays are filled they can be stored as shown until ready to be transferred to the incubator

must not be set.

Hatching eggs should be collected three or even four times a day, for the sooner the internal temperature can be brought down to 15-16°C the better will be the hatching results. Eggs should be collected in containers which allow them to cool quickly.

Dirty eggs should be dry cleaned with wire wool. If contamination on the shell is extensive there may be no alternative but to wash it. In this case the water temperature should be around 30°C and should contain a germicidal compound. After washing, the eggs should be carefully dried before storage.

The best temperature for storage is between 10-15°C. The relative humidity should be between 75 and 85%. Hatching eggs should not be stored for longer than 7 days; after this period the hatchability rapidly declines. The eggs should always be handled with extreme care and not jarred or shaken, for this may damage the contents. Fig. 2.2 shows the eggs placed in the setting trays.

PHYSICAL REQUISITES FOR SUCCESSFUL INCUBATION

The one golden rule in artificial incubation is: Carry out the operational instructions advised by the manufacturer of the incubator.

Temperature

Normal hen temperature is between 40.5°C and 41.7°C. The optimum temperature for hatching is 37.7°C at the centre of the egg. With broody hens the temperature is identical, but the surface temperature of the egg may read 39.3°C. With flat-type or still-air incubators the temperature reading is taken on the surface of the eggs and should be 39.3°C. Forced-draught incubators or cabinet incubators, on the other hand, force or blow warm air round the egg, and a reading of 37.7°C is taken as there is no variation between internal or surface temperature. When separate hatchers are used the temperature may read 37.2°C or 37.5°C. This is because considerable heat is produced by the developing embryo during the last 2 days of incubation.

Relative humidity

The relative humidity within the incubator can fluctuate far more without seriously affecting hatchability than can the temperature. There is, however, an optimum value.

Relative humidity can be measured by comparison of dry-bulb and

Top: Anconas: Fine layers of white eggs

Bottom: White Leghorns. Another Mediterranean breed; good layers and white eggs. Not as tame as heavier breeds.

Andalusians: Sound layers of white eggs. Do not come broody. Quite scarce, but well worth keeping for their beauty and laying. Too small for table birds. May be on the wild side for small garden.

Old English Game: Black Red Cock and Wheaten hen. Rather dark wheaten by today's standards–more like the 'Clay' colour. Reasonable layers and fine table birds.

Light Brahmas: Large, exotic birds; moderate layers of tinted eggs. Feathered legs and Pea comb make them very attractive. Heavy breed which will not fly over fences or become wild in behaviour.

Dark Brahmas. Important because they were the breed used to create others. The dark colour helps to keep them clean.

Top: Campines. These days a Hennie type of cock is preferred (with no sickles). Lays about 150 white eggs a year.

Bottom: Dorkings – Silver cock and hens, and dark hen. Heavy bird, but no longer very productive. Five toes.

Buff Cochin: Only fair layers. Very feathery and majestic. Not good layers; tinted eggs.

Partridge Cochin: Very attractive colour, but not a commercial type bird. Single combs, whereas Brahmas have a Pea Comb.

Top: **White Dorkings which have Rose combs, and Red Dorkings which are quite rare. One of the oldest breeds, but have lost ground.**

Bottom: **Faverolles: French breed for laying and table. Tinted eggs in reasonable numbers. The colour shown is really a non-standard variety because the nearest is the Salmon which is similar to the above, but has shoulders of cherry red and gold (USA Reddish brown). Note the muffles, five toes and feathered legs. Weight of cock about 5 kilos (11lb.).**

Top: Gold Pencilled Hamburghs. Once very popular and quite beautiful.

Bottom: Silver Spangled Hamburghs. Spangles and tips make this a very attractive breed. Light Breed.
Both lay white eggs in reasonable quantities, but on the small size.

Top: Houdans. Attractive with crest and leaf comb, but only fair layer.

Bottom: La Fleche. French like Houdan. Lays a white egg and is a good table bird, but rarely seen. Unusual horn comb.

Croad Langshan. Good layers of deep brown eggs as well as being commendable table birds. Probably source of deep brown eggs for others dark brown egg layers. ('Coffee' Coloured)

Modern Langshan. Taller than *Croad* on which it was based. Brilliant black plumage, but now rare.

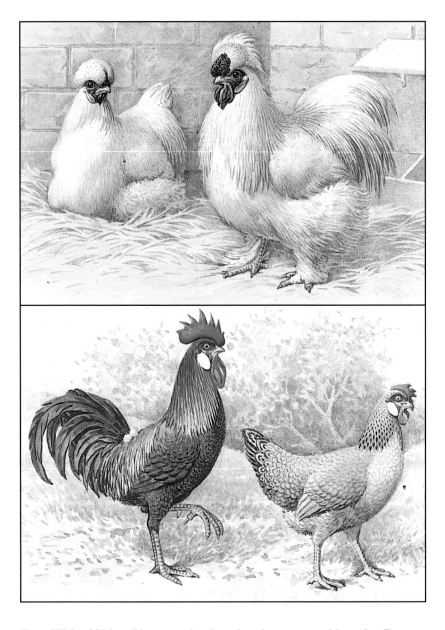

Top: White Silkies. Very popular breed and a renowned broody. Become very tame. A good layer when not broody, but not a table bird.

Bottom: Brown Leghorns. Very attractive member of white egg layers.

Top: White Minorcas: Delightful member of Minorca family. Many prefer Black variety because they keep cleaner looking. Non-sitter. Quite rare along with Blues.

Bottom: Black Minorcas which are great favourites.

Top: Buff Orpingtons. A great favourite in days gone by. Dual purpose breed for laying and for the table. Laying abilities of some strains rather poor because they are now too feathery.

Bottom: Blue Orpingtons. Original type as developed by William Cook. Very popular and quite attractive. Large birds for table.

Top: Barred Plymouth Rocks. A popular dual purpose, heavy breed which lays well (tinted eggs). Very old breed which settles well on free range or in a small area.

Bottom: White Wyandottes. Like the Rocks is an American breed which lays well. Heavy breed which lays tinted eggs. Suitable for free range or semi-intensive in open run.

Top: Dumpies or Scots Dumpies. Dwarf-like breed which lays reasonably well, but belongs to a special race known as "Creepers" which, due to the genetic makeup may carry a lethal gene. Lays white eggs. Light breed.

Bottom: Scots Grey: A gamey type breed which is barred. Good layer of white eggs and quite attractive, although not many are kept. Non-sitter.

Red Sussex. Dual purpose breed which lays light brown eggs. Fattens well for the table. Mature cock over 4 kilos (about 9lb.) This variety rare.

Light Sussex. These are a utility type. They are very popular, but for showing the hackle and tail must have very dark markings.

Various Breeds

Barnevelders, Marans, Welsummers (Deep brown eggs)
Rhode Island Reds (Brown eggs), White Leghorns (White eggs),
Indian Game (USA Cornish) Table bird, especially when
crossed with white skinned breed.

wet-bulb readings. At the normal temperatures and R.H. recommended for cabinet incubators the dry-bulb should read 37.7°C and the wet-bulb 31°C. A low R.H. results in too much moisture loss from the egg; hatching will be delayed and the results poor. Too high R.H. results in wet sticky chicks, and many drown in the egg. Incubators with separate hatchers usually have a lowered R.H. from the 19th day of incubation. The wet-bulb should be reduced to 24°C for 36 hours. It is then raised to 32.2°C until 2 hours or so before hatching, when it is again reduced to 29.4°C.

Ventilation

The developing embryo requires oxygen. Carbon dioxide and other gases must be removed from the egg. Ventilation therefore serves two purposes: one to supply oxygen, the other to remove harmful gases. For this reason, the incubator room must be well ventilated, since it is from this source that the incubator draws its air.

Position of the eggs during incubation

The ideal position of the egg for incubation is either flat on its side in natural-draught incubators or with its broad end uppermost in forced-draught machines. In these positions warm air is allowed to circulate freely, and the embryo can orientate itself satisfactorily. Facing the narrow end uppermost causes many embryos to adopt mal-positions and also fail to penetrate the air space at hatching, which allows pulmonary respiration to commence.

Turning eggs

The more eggs are turned during incubation the better, usually, is the hatch. Turning prevents embryonic membranes from sticking together, allows correct orientation of the embryo, ensures an even distribution of air and prevents the embryo from sticking to the side of the egg.

The needs for turning are greatest during the early stages of incubation and decrease gradually up to the 19th day. After this, turning should cease. The number of times the eggs are turned varies from every hour to five times a day. The latter number usually applies to natural-draught incubators which are turned by hand. It is important to turn the eggs an odd number of times each day. By so doing, the embryo does not spend the long unturned period on the same side each night. With mechanical turning this is, of course, not important.

Natural incubation

Hatching under broody hens is ideal for the small man raising only a dozen or so chicks each year. The number of hen eggs one broody can incubate varies between 12 and 15. The nest box should have a base 35 cm x 35 cm and be 40 cm high. The centre of the nest should be saucer shaped to prevent eggs from rolling out. Hay or straw may be used as nesting materials. Before setting the broody hen it should be dusted with insecticide to free it from external parasites. It should be allowed off the nest once a day for feeding, watering and emptying its bowels, it should not leave the nest for periods longer than 20 minutes, depending upon weather, etc. As soon as the hatch is off, all debris and unhatched eggs should be removed. Fresh litter should be given and the hen and chicks moved to clean quarters.

ARTIFICIAL INCUBATION

Incubators are of two main types -

1. Natural draught **and   2.** Forced draught.

The *Natural Draught* (Still Air) are **small machines** with eggs on only one level. The air passes through the machine and around the eggs by natural circulation. Modern machines now have sophisticated electronic controls so are more efficient than the older machines.

Forced draught machines are usually large with egg trays one above the other, and the air is forced through by a fan. These are known as "cabinet" machines. Both hatch equally well if well managed, but the cabinet machines need less labour so that most larger poultry farms now use them.

There is a limit to the size or number of eggs the natural draught machines will take, from 100 to 150 eggs being the most satisfactory size.   Large machines do not hatch so well.   The cabinet machine is almost unlimited in size, some taking many thousands of eggs. From 1,000 to 20,000 eggs are the most popular sizes for poultry farmers.

TYPE OF MACHINE TO BUY

For farms with under 300 head of breeding stock it is possibly best to have natural draught machines. For 300 breeders or more, a cabinet will save labour. With natural draught machines, the size stated, such as 150 eggs,will take this number and hatch them out in the one compartment. If it is necessary to hatch 50 eggs a week, four 50-egg incubators will be needed, as it takes about a week to finish off the hatch and clean up for the next one.

To do the same work a cabinet would have to have a capacity of 200 eggs. Such small ones are not made, but this is mentioned because the capacity of a cabinet machine will take, each week, a fourth of its rated capacity, i.e. a 2,000 egg machine will take 500 eggs a week.

A natural draught machine hatches the chicks in the same tray as the eggs are set. In a cabinet machine the eggs are transferred to special hatching trays on the 18th day of incubation.

When estimating the incubator capacity required it should be borne in mind that about three eggs have to be set to produce one pullet chick for certain. If, therefore, an output of, say, 300 pullets a week is required a machine or machines capable of taking about 1,000 eggs will be necessary.

LARGE INCUBATORS

The large cabinet incubators require a different style of management because thousands of eggs are being set and , as a result, mistakes can be very costly.Moreover, the large scale operation will need watchfulness on all costs and especially utilization.

Setters & Hatchers

Modern practice favours the use of one or more **setters** and an appropriate number of **hatchers**. On the eighteenth day, the eggs are transferred to the hatcher which can be given an uplift in temperature to hatch the chicks (depends on the exact type of machine).

An example is given on the next page and the two machines are similar in layout, but the conditions at the end are different, thus maximizing the hatch.

A Setting Table

With the very large machines a special setting table may be used thus allowing large quantities to be handled quickly and safely. A vacuum egg lifter may be used with air being supplied from a compressor. On one system the manufacturers claim an egg setting speed of 30-35,000 eggs per hour with a labour utilization of four personnel on the operation. An example is shown overleaf.

The trays are mounted on a trolley which is pushed into the cabinet machine when full to the required capacity.

Egg Setting Table
Speedy setting by ease of lifting on to trays

Setter for Hatching

Similar to the Hatcher with identical trays which can be moved from one machine to the other. (Courtesy: Petersime).

Walk-in Systems
More modern incubators, especially in hatcheries with large outlets, are of the walk-in type. This means that the operator can comfortably walk into the machine and attend to his work without stooping, etc. The main advantage of walk-in incubators is the great economy of space. The walls and ceilings are so constructed that the surface is easily cleaned and able to stand up to strong disinfection. Insulation must be of the highest order.

In design of hatcheries great care must be taken to lay out a system which will be efficient to work and at the same time be simple to disinfect. Provision should be made for an egg holding room of temperature 12.8-15.5^{0}C, a prefumigating room in which eggs can be fumigated prior to setting, a sexing room and, if necessary, a packing and despatch room.

The ventilation of the incubator room should be set to change the air every 8 hours. Air inlets should be low down and baffled to prevent draughts. Outlets should be placed in the roof of the building and, far preference, should be fan assisted.

HATCHING HYGIENE
The hatchery should be kept as clean as a maternity hospital. It merits the most meticulous attention to hygiene at every stage of its operation.

Infection may enter the hatchery by four main routes:

(a) Surface of eggs and collection containers

(b) Vermin and ecto-parasites

(c) Flies

(d) Hatchery operators

Visitors should be kept out of the hatchery and staff should wash their hands periodically throughout the day. Sexers who visit the hatchery should disinfect themselves and their clothing before entering or leaving the hatchery. Chick boxes must be of a non-returnable type. The use of aerosol sprays in the hatchery is to be recommended. They keep down the bacteria count in the building but they do not replace disinfection.

Egg setting and hatching trays should be steam cleaned with boiling water containing a 4% solution of washing soda. Internal incubator fitments, walls, sides and floors should be treated in the same way.

Fumigation can be carried out by using potassium permanganate and formalin. When mixed, these two substances react violently with the liberation of formaldehyde gas. Rubber gloves must be worn when disinfecting with these two chemicals, for injury to the operator is not uncommon when negligence is evident. The amount of these two chemicals to use depends upon the size of incubators and desired strength. For each 2.83m^3 of incubator space the following is recommended:

Formalin 128 ml

Potassium permanganate 85 ml

This strength is used in the setting compartment. Combined setters and hatchers should be fumigated for 30 minutes at the beginning of each season, with all ventilation ports closed. After each loading of eggs fumigation should be carried out for 30 minutes.

VENT – SEXING

Although a number of chicks are sold 'as hatched' an increasing number are sold already sexed. This is usually carried out by men expert in their job. When it is the intention of using the services of a sexer the following equipment will be necessary:

(a) A bright lamp with a reflector to shade the light from the sexer's eyes and direct it on to the chicks

(b) An enamel bucket for the faeces

(c) Two chick boxes: one for pullets and one for cockerels.

The sexer first evacuates the faeces from the chick's intestines by gentle pressure on the abdomen. The chick's cloaca is then everted and examined under the bright sexing lamp. By recognition of the variation between the eminence of pullet and cockerels the chicks are sexed. The usual degree of accuracy is 98% and the speed of sexing is about 1,000 chicks an hour. Great skill is required to reach these standards.

Sexing by machine was popular about 15 years agso, but since then it has declined in popularity. That this has happened is mainly due to lack of speed achieved by this method. However, the training required to obtain reasonable accuracy is small in comparison with hand sexing.

The machine consists of a hollow tube in which a light is reflected. This is inserted into the chick's body. The operator, looking through an eyepiece, is able to recognise or distinguish between the male and female sex by the presence of testes or ovary.

Sexing by machine. The sex organs are illuminated by a light reflected back into the eye piece

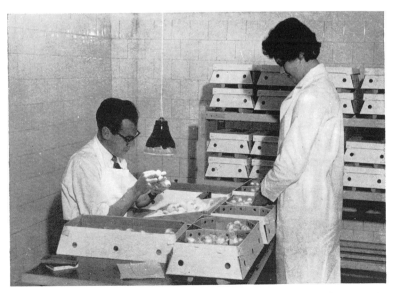

Manual method of sexing day-old chicks. This method was perfected in Japan and in skilled hands great accuracy at high speed can be attained.

Testing for fertility

The testing of eggs in natural-draught incubators is carried out on the 14th day of incubation. In cabinet or forced-draught incubators it is performed on the 18th day before moving the eggs to the hatcher or, hatching compartment. Testing is done to determine the number of eggs which are fertile or which are alive. Infertile eggs are removed from the trays and usually sold to the catering and bakery trade.

Testing is carried out by placing the eggs over a strong light which shines through them. They can be 'candled' individually or *en bloc*. Infertile eggs are seen as 'clear' (that is, the light shines through them), but with a fertile egg part of it is dense in appearance. Not all apparently 'clear' eggs are infertile: some contain embryos which have died at an early stage of development. These can be determined by breaking out a number of 'clears'.

Percentage fertility is defined as:

$$\frac{\text{Number of eggs fertile}}{\text{Number of eggs set}} \times 100$$

COMMON REASONS FOR FAILURES IN INCUBATION

Trouble	*Probable reason*
Delayed hatch	Setting stale eggs; low incubator temperature
Early hatch	Temperature too high
Many dead in shell	Incorrect turning of eggs; faulty breeding stock; inadequate nutrition; Salmonellae infection.
Weak chicks	Overheating of hatching unit
Sticky chicks	Temperature too high
Small chicks	Insufficient humidity; setting small eggs
Too many infertiles	Insufficient males in breeding pen; males too old; mismanagement of cockerels; eggs kept too long under adverse storage conditions.

FACTORS AFFECTING HATCHABILITY

Nutrition

Mortality of embryos is normally associated with the middle term of development i.e. $10 - 14$ days of incubation for hen eggs. Breeding stock must be fed rations rich in vitamins and trace elements.

Breeder flock age

The older the flock the lower the hatchability. Fertility decreases with flock age.

Pre-incubation treatment

Dirty eggs hatch less well than clean ones and naturally clean eggs better than those artificially cleaned. Clean nest boxes and clean nest litter produce clean eggs. Frequent egg collection is important.

MARKING DAY-OLD CHICKS

Day-old chicks may be marked in the following ways for ease of identification:

Toe-punching. A small hole is made in the web between the chick's toes. There are no fewer than 16 combinations.

Leg ringing. A ring is attached just above the chick's foot. A considerable amount of work is involved, as they need frequent changing as the bird grows.

Wing banding or tabbing. The chick is marked by attaching an aluminium tab to the web of its wing. If fitted correctly it should never be lost. The flap of skin at the front of the chick's 'elbow' is used for inserting the pin of the tab. No pain is caused the chick.

Dyes. Dyes may be used to temporarily mark baby chicks. They wear off as feather growth progresses. The usual site for marking is on the chick's back.

Chapter 3

CHICK BROODING AND REARING

INTRODUCTION

Baby chicks, unlike most young animals, are unable to live for long in the normal temperature, but after a short time outside must be able to go somewhere to warm up. They can stand quite low temperatures, even freezing, if they can get into the warmth as soon as they feel cold. If they are out too long, even in temperatures which appear quite warm to human beings, they become chilled and will die or grow stunted. A slightly chilled chick does not die at once, but digestive troubles commence and the chick may die in about a week.

Rearing hardy stock depends on the combination of three factors. These are:

  (a) Quality of stock
  (b) Good management
  (c) Correct nutrition

It is false economy to buy the cheapest stock. Good quality stock, from a reputable breeder, is worth any additional expense. Those which appear chilled or are ailing must not be accepted in the hope that they will revive.

Good management is essential for success to be achieved. This involves a certain amount of 'stockmanship' and, above all, commonsense. No bird is so good that it cannot be improved by good nutrition or so poor that it cannot deteriorate with poor nutrition. Good nutrition is essential during the bird's first few weeks of life. As good consumption is low during the first 6 weeks food cost per ton should not be used as the main criterion for choosing a suitable ration.

Date of hatching

The light breeds, light crosses and hybrids used in commercial egg production do not take as long to reach sexual and physical maturity as the heavy breeds, heavy crosses and hybrids. Some light-type chickens may reach sexual maturity by 4 months of age, while some heavy types may take 2 months longer.

Rate of sexual maturity is governed not only by genetic factors but also by the date of hatching. Chicks hatched during the winter and spring months reach sexual maturity about a fortnight before chicks hatched during the summer months. This is because increasing and decreasing natural daylengths affect hormone production, which governs age of sexual maturity. Chicks reared with increasing daylength mature earlier than those reared on decreasing daylength.

Bearing the above facts in mind, it is possible to programme a rearing pattern for all-the-round rearing and egg production. Winter rearing is difficult under extensive conditions, but intensive methods are practised with complete satisfaction.

Autumn chick rearing is popular. This is because chicks hatched at this period of the year reach sexual maturity and commence laying at a profitable time. Spring hatching and rearing is, however, still popular, for it makes use of less intensive management systems under better natural environmental conditions. It also fits into the all-the-year-round rearing and egg production scheme.

NATURAL BROODING

The period from hatching to that when the chicks no longer require heat is known as the *brooding period*. It usually varies from 5 to 8 weeks. Hens may be used to brood and rear small groups of chicks. An average-size hen will brood from 10 to 15 chicks, depending upon the weather. Brooding is usually carried out in a hen coop which measures 0.6 m x 0.6 x 0.6 m. It must be weather – and vermin-proof. The hen is confined to the coop by bars, 8 cm apart. These allow the chicks out of the coop but restrain the hen. Litter, peat moss or wood shavings should be used on the floor of the brooding coop. Hen brooding is only practised when small numbers of chicks are reared. It is not practical or economic with large poultry units.

Broody hens cannot be bettered for raising chicks, and often are still used where small numbers are involved. Even when it has not hatched them, a broody can be made to take chicks, provided they are carefully introduced.

Range shelters are excellent for rearing healthy stock

ARTIFICIAL BROODING

Artificial brooding infers the use of equipment which provides conditions similar to those of the brooding hen but in much larger numbers.

Temperature Requirement

The optimum brooding temperature is between 35°C and 37.7°C measured at a height 5 cm above floor level. As the chicks grow and feather so the temperature requirement decreases. The following is a guide, according to weeks of age, for brooding temperatures:

Age in weeks	Temperature in °C
0 – 1	32 – 35
1 – 2	30 – 32
2 – 3	27 – 30
3 – 4	24 – 27
4 – 5	21 – 24
5 – 6	18 – 21
6 – 7	15 – 18

The best guide to temperature requirement is the state of the chicks. If they are huddled around the heating unit the temperature is too low; if they are widely spread out the temperature is too high. Chicks which are too hot or cold cheep plaintively. Contented chicks are those which are evenly spread out over the brooding area, cheeping only occasionally.

Manufacturers of brooding equipment usually provide instructions. These should be followed unless proved incorrect.

Sources of heat

Heat for brooding can be supplied by electricity, gas, solid fuel, paraffin oil or hot water. Electricity is easy to operate and is clean, but in some areas the source is unreliable owing to voltage drops, power failures or breakdown of equipment. The greatest benefit is that it is clean and easily controlled. There are many types of electric brooders on the market, varying from infra-red lamps to heat storage brooders to large canopy types.

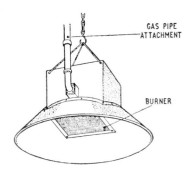

Fig. 3.1 Gas infra-red brooder,
28 cm diameter by 12 cm high,
suitable for 500 chicks or 250
turkey poults.

GAS PIPE
ATTACHMENT

BURNER

Infra-red brooders (Fig. 3.1.) are cheap and efficient to use, requiring a minumum of space, and there are no fumes or canopies to restrict ventilation. Adjustment to temperature is carried out by raising or lowering the infra-red lamp, and the chicks are always visible to the operator. The units are in two forms: dull and bright emitters. Surrounds for the brooder should be 50 cm high to cut out floor draughts.

In the summer months the brooders can be turned off during the day when the chicks are 3 weeks old. The lamps are usually of the 250 watt type, which is suitable for 75 chicks from 1 day old to 6 weeks of age. Where larger groups are brooded several lamps may be wired together and made to form a triangle or fixed in line side by side. Paraffin infra-red brooders are also obtainable for small scale rearing (Fig. 3.2). Food should not be placed under the lamp rays.

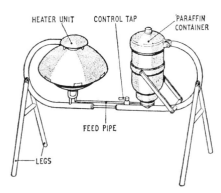

HEATER UNIT   CONTROL TAP   PARAFFIN
CONTAINER

FEED PIPE

Fig. 3.2. Paraffin heated infra-red brooder. Burns 2.27 litres paraffin each 30 hours. Capacity 200 chicks or 100 turkey poults.

LEGS

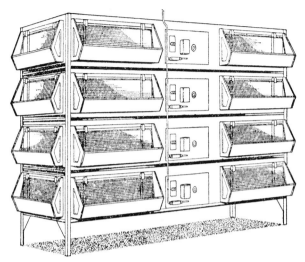

Fig. 3.3. Gas tier brooders 2.5 m x 0.9 m per tier,
150 chicks per tier from day old to 4 weeks

Tier brooders

This type of brooding is used at $3 - 4$ weeks of age. After this the chicks are transferred either to haybox brooders or into intensive rearing houses. About $0.093m^2$ is allowed to every 6 chicks. The units are generally in blocks of 5 tiers (Fig. 3.3) and heated by electricity, paraffin or gas, the heating pad being placed centrally or to one end of the unit. Usually 80 chicks are brooded in each tier; thus a block of 5 tiers has a capacity of 400 chicks to $3 - 4$ weeks of age. This brooding system saves a considerable amount of space compared with floor rearing methods. Food and water are provided in troughs attached to the side and ends of the unit. The building housing the brooder should be heated to $18^oC - 21^oC$.

Heated floor brooding

Heated pipes or wires are positioned under or in the brooder house floor, being built into concrete or laid in sand. Installation costs are high, but when large numbers of chicks are brooded the running costs are low. The system has the advantage of being eligible for lower running costs as it can be used in off-peak hours as an electric heat-storage brooder. The chicks have little chance of being chilled as the floor is

A wire platform is a useful addition to pyramid hovers after the first week. It is best, however, not to use this platform in the first week of brooding as chilling may take place in the colder weather

Intensive brooding using the infra-red system. The lamp is raised 5 cm each week

usually evenly heated over its area.

Maintenance costs can be high in the event of a fault under-floor.

Hovers

A hover consists of a pyramidical metal canopy mounted on legs under which a heating unit is placed. The brooding system is useful for both the small poultry or large poultryman. This type of brooder has to be housed in a brooder house. The chicks are retained by metal sides or felt curtains. Heating is by oil, electricity or gas. Fig. 3.4 shows a pyramid hover suitable for the small flock.

Many types of paraffin-gas brooders are on the market and do excellent jobs. A number of them are manufactured for use in mass brooding methods, such as in broiler production. They are usually suspended from the building roof. Some have a canopy, others do not. Those without canopies are cheaper to purchase but can only be used in well-insulated buildings, as much of the heat is used for space warming. Brooders with canopies should be used where the insulation of the building is inadequate.

Fig. 3.4. One of the popular pyramid hovers. In this case heat is provided by electricity, and there is an inspection window in the canopy, while the sides have detachable metal shutters.

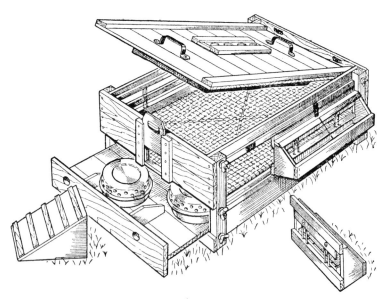

Fig. 3.5. A warm floor brooder. Heat is provided by lamps under a wire floor. Detachable food and water troughs are placed along each side, and they can be adjusted as the chicks grow.

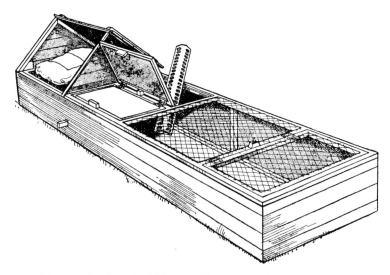

Fig. 3.6. A hay box brooder fold. Bags of hay are placed on a wire frame in the end section, and the food and water troughs are under a covered portion of the run

Warm-floor brooders

This type of brooder is essentially for the small flock owner. It consists of a unit 1 m x 0.65 m with boarded sides and ends. Food and water troughs are mounted on each side. A pophole is fitted to one or both ends. The floor consists of 1.25 cm wire mesh over which a blanket-type canopy is fitted. The heater unit is fitted in a drawer under the wire floor. Small paraffin 'putnam' stoves are used. The heating is regulated by addition or removal of these lamps. Electrically operated units are also available. This type of unit, shown in Fig. 3.5 can be used for intensive or extensive rearing of chicks. To avoid overcrowding the numbers of chicks should be reduced by half at 2 weeks of age.

Range brooding

Despite the popularity of intensive rearing, range rearing is still carried out by some poultry farmers, essentially the smaller enterprises. There is no doubt that during the spring and summer months chicks benefit greatly from being reared on good pasture. Feathering is generally better compared with intensive rearing. Chicks are generally brooded intensively for the first 4 weeks by one of the methods described earlier. After 4 weeks hay box brooding may be practised. A hay box measures 1 m wide by 3 m long. One end, 1 m x 1 m, consists of a brooding compartment in which hay bags are used to conserve body heat. Each unit had a rearing capacity of 33 – 35 chicks to 8 weeks of age. Feed hoppers are fitted to the brooding square and also in the run unit. Hay box fold brooders (Fig. 3.6) should be moved daily to prevent 'poaching' of the ground. Level ground is essential to prevent the chicks from escaping.

After 8 weeks of age the growing stock may be transferred to night arks or range shelters. Night arks are 2 m long by 1 m wide. Each one holds 25 growing pullets from 8 weeks to point of lay. They should be spaced 3 m apart and moved occasionally to avoid 'burning' the ground.

Range shelters consist of a series of perches under a low apex roof. The sides and ends are covered with 2.5 cm mesh. They are light and therefore easily moved. Each range shelter holds 50 growing pullets to point of lay. They should be spaced about 20 m apart. These shelters can also be used for rearing turkey breeding stock to maturity.

Battery of tier brooders are commonly used in the early stages of brooding, particularly for table birds

Large scale pullet rearing on litter

Chick Brooding and Rearing

Rearing on wire or slatted floors

Many large-scale rearers of growing stock house their pullets under extremely intensive conditions. The pullets are reared to 16 — 18 weeks of age in houses similar to those used for broiler production. The stocking intensity is 0.093 m^2 per bird up to point of lay. Food is usually provided by automatic floor feeders and water troughs run down either side and middle of the house. Lighting is completely artificial, and ventilation is regulated by thermostatically controlled fan extractor units. A high degree of management is essential to prevent smothering. Culling is virtually impossible under the intensive conditions but removal of obviously sick birds is essential.

MANAGEMENT OF GROWING BIRDS

The house used for rearing must be warm and well ventilated. It should be dry and free from holes and cracks. The brooder should be heated for 2 or 3 days before the chicks are due, making certain that the required temperature is achieved. Corners into which chicks or growers may huddle and crowd should be rounded off to deter smothering. Sufficient food and water troughs must be provided and placed so that the chicks can feed and drink in comfort. Water must not be too deep or the chicks may drown.

Floor space recommendations

System	Age (weeks)	Floor space in square metres/100 birds
Floor/litter	0 — 4	4.0
	4 — 8	9.0
	8 — 20	12.0
Cages	0 — 4	1.0
	4 — 8	3.0
	8 — 20	4.0

Daily Water consumption guide/100 birds

Day old to 7 days	0.4 litres
7 — 30 days	1.0 litres
30 — 56 days	1.75 litres
56 — 100 days	2.25 litres

Food consumption

The following is a guide to the cumulative food consumption in kilograms per 100 pullets for birds fed ad libitum.

Age in weeks	Light type	Heavy type
1	9.00	11.5
2	18.00	22.0
3	27.3	45.0
4	45.0	68.0
5	68.0	113.0
6	90.0	160.0
7	136.0	210.0
8	193.0	250.0
9	210.0	295.0
10	250.0	354.0
11	286.0	410.0
12	341.0	464.0
13	395.0	545.0
14	454.0	613.0
15	509.0	680.0
16	568.0	750.0
17	613.0	804.0
18	680.0	873.0
19	750.0	954.0
20	829.0	1000.0
21	863.0	1045.0
22	918.0	1070.0
23	954.0	1090.0
24	1000.0	1140.0

Dubbing and de-spurring

Breeding replacement stock may have the comb dubbed. Where no importance is attached to comb and wattle growth they can be removed when the birds are day old. Dubbing prevents damage to those parts when large combed varieties are housed in battery cages. Scissors are used for dubbing. The comb is removed close to the bird's head cutting from front to rear. De-spurring in male stock helps prevent damage to females at mating. Spurs are removed by equipment used to debeak birds.

Most light breeds and hybrids can be separated by 6 − 8 weeks of age. Heavier types can be sexed by 9 − 10 weeks, sometimes earlier. The sexing is done by comb and wattle differentiation, feather growth and body conformation.

The comb and wattles in male birds are larger than those of pullets. The feathers on the male's back are pointed, while those of the female are rounded. Tail feathers in males are usually but not always less well developed than females. The body of the male is deep and square. The female's body is longer and shallower.

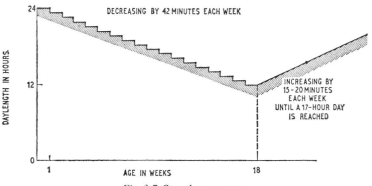

Fig. 3.7. Step-down system.

LIGHTING THE GROWING PULLET

It was mentioned at the beginning of this chapter that decreasing day-lengths delayed the onset of sexual maturity and an increasing one encourages its onset. Bearing these facts in mind a system of lighting has been developed called the 'step-down-step-up' system. It has been

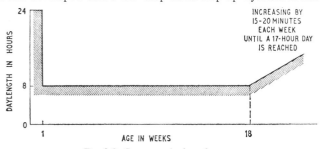

Fig. 3.8. Constant daylength system

developed to counteract the stimulating effect of increasing hours of natural daylight. Chicks hatched in the period October to February, when reared in windowed rearing houses, mature too early. The 'step-down' lighting system counteracts this by providing a long artificial daylength at the commencement of the chick's life. The amount of light is decreased weekly until the amount of artificial light and natural daylight coincide. For obvious reasons no further decreases are possible. As it is possible to calculate the weekly reduction the period of equality between natural and artificial daylength can be made to occur at point of lay.

The popular 6ft by 4ft house, suitable for a great variety of purposes. With a hover it can be used as a brooding unit for chicks, or it can accommodate growers, cockerels or a small flock of layers

When it is desired to apply the 'step-down' lighting system the following points must be known:
(a) Genetic age of point of lay, e.g. 18 weeks
(b) Length of natural daylength at point of lay, e.g. 12 hours
(c) Delay in sexual maturity required, e.g. 2 weeks

The Sussex ark is very useful for rearing, but it has also many other applications

Knowing these points the following procedure may be followed. From day-old to 1 week provide 24 hours of light. From 1 week to 18 weeks (i.e. for 17 weeks) decrease the amount of light in equal weekly amounts e.g. from 24 hours to 12 hours, a total reduction of 12 hours. The rate of reduction is therefore 12 hours in 17 weeks i.e. 42 minutes a week (See Fig. 3.7).

The main benefit to be gained from using this lighting system is the increased initial egg size. This is because the 2 weeks' delay in sexual maturity has allowed greater body development. Egg production is not necessarily increased with this lighting system.

Poultry keepers rearing their pullets in windowless, controlled environment houses may use another lighting system. This is known as a 'constant daylength' lighting pattern. It does not delay sexual maturity by more than a few days, but, because of the increased lighting during the laying period, improves total annual egg production compared to other systems.

The system is as follows. The pullets are started on 24 hours of light. This is maintained for 1 week. From 1 week to 18 or 20 weeks the amount of light is kept constant at 6 or 8 hours (see Fig. 3.8).

Modifications of these two lighting systems are used in practice. All, however, set out to achieve the same results of either improved egg size, egg production or both.

PULLETS AT POINT OF LAY

As a pullet approaches the point of lay stage its body undergoes great physiological changes. The ovary becomes increasingly active from about 16 weeks of age; parallel with this the bird's oviduct is preparing itself to receive the ova. Added to this must be the 'stress' effects of handling, debeaking, vaccination, housing, feed changes and competition from other birds. The net effect of all these changes in management is to reduce the bird's resistance to disease. It is therefore vitally important that the bird should be handled most carefully at this time. Food changes must be made gradually to keep the effects on stock down to a minimum. The change from the growers to the layers rations should be made 2-weeks before egg production is expected. An 18-week change is normal.

Chapter 4

NUTRITION AND FEEDING

INTRODUCTION

Food is of no value whatsoever until it has been eaten, digested and absorbed, and therefore a brief knowledge of these processes is essential to the correct understanding of nutrition.

The digestive system of the fowl is shown in Fig. 4.1. The fowl has no teeth but possesses a hard, horny beak. Food picked up and swallowed passes quickly into the gullet and oesophagus, which leads into the crop. In the crop the food is moistened and prepared for digestion; the prime function of the crop is storage. Connecting the crop to the proventriculus or glandual stomach is a short tube. In the proventriculus food is acted upon by enzymes and hydrochloric acid. The gizzard lies just beyond the proventriculus. It is a highly muscular organ and crushes the food by pressure into a cream-like pulp before passing it into the first part of the small intestine, the duodenal loop. Within the confines of this loop lies the pancreas, which liberates pancreatic juice. The liver communicates with the small intestine by two ducts, one of which is connected to the gall bladder, which stores the bile salts.

Lining the inner surface of the small intestine are innumerable, small finger-like projections called *villi*. These increase the area for absorption, and each is connected to the blood system, which carries the absorbed food to the parts of the body where it is needed. At the junction of the small and large intestine is a valve which prevents the return of food contents. Arising from this junction are two blind pouches known as caeca. Their function is to digest fibre and absorb water. The caeca empty at eight-hourly intervals. The alimentary canal terminates in the cloaca.

The fuel which the bird's body burns is food. It enables the vital processes to be carried out and provides material for maintenance, production and reproduction. The essential nutrients which the ration must provide are:

(a) Water
(b) Energy
(c) Protein
(d) Fats and Oils
(e) Vitamins
(f) Minerals

Water

Water is an essential part of the ration. The body contains 60% and the egg 65% water. Lack of it seriously retards growth and production, and absence of it leads to death in a short period of time. Water is essential for the absorption of essential nutrients. It helps maintain body

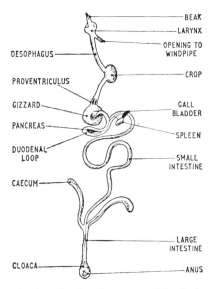

Fig. 4.1. The digestive system of the fowl

54

temperature and is essential for removing toxic products from the bird's kidneys. Water also acts as a lubricant for bone joints and maintains the blood in its proper state of consistency. It is the simplest of all compounds employed in nutrition, the easiest to supply and also the cheapest.

A water consumption guide is given below:

Age in weeks	Gals/ 100 birds/day	Litres	Age in weeks	Gals/ 100 birds/day	Litres
1	$^2/_3$	1.49	11	3 $^1/_2$	15.90
2	$^1/_2$	2.27	12	4	18.18
3	$^3/_4$	3.76	13	4 $^1/_4$	19.31
4	1 $^1/_4$	5.68	14	4 $^1/_4$	19.31
5	1 $^3/_4$	7.95	15	4 $^1/_2$	20.45
6	2	9.09	16	4 $^1/_2$	20.45
7	2 $^1/_4$	10.22	17	4 $^3/_4$	21.58
8	2 $^1/_2$	11.36	18	4 $^3/_4$	21.58
9	2 $^3/_4$	12.50	19	5	22.72
10	3	13.63	20	5	22.72

Energy

This covers a large class of compounds, including the fats and oils dealt with later. Included under this heading are sugars, starches, cellulose and fibres. The energy portion of the fowl's ration constitutes a very large portion. Energy provides heat to do work and is found in all the cereal grains in varying amounts. The richest sources of energy are the fats and oils, providing 2¼ times the amount provided by the cereals. Of the cereals, maize, milo or sorghum and wheat are the main energy suppliers. Energy is uaually expressed as metabolisable energy (M.E.) in kcals/kg and MJ/kg. (Mega Joules/kilogram).

The M.E. of a food is the amount of digestible energy in the food less the amount of energy in the faeces. Productive energy is the net energy less the energy needed for maintenance. It is the net energy stored up as fat and protein in a growing or fattening bird. Foods or ingredients with high-energy contents usually have low fibre values. Thus, a food such as maize with a M.E. energy level of 3390 kcals/kg has a fibre content of 3%. Fibre is of very little value to poultry because food passes quickly through the digestive tract, and so bacteria do not play an appreciable part in digestion, as they do in other animals.

The higher the fibre content of the ration the lower its energy value, and if fibre is fed at a high level it may be impossible for the birds to consume sufficient food to obtain adequate nourishment. Birds eat primarily to satisfy their energy needs. Thus a ration containing high energy is consumed in smaller quantity than one containing low energy.

Protein

Proteins are an essential constituent of muscles, blood and feathers. They are extremely complex substances made up of amino acids. These amino acids, in their correct ratio, are built up by the bird into muscle, egg or feather protein. Excess protein is broken down, some being used for energy purposes, the remainder being excreted via the faeces. Amino acids are either essential or non-essential. Essential ones are those which have to be provided by the ration in the correct ratios, as a deficiency of any one will adversely affect growth rate or egg production. As the bird has a limited ability to store protein a continuous source must be available. The amount of amino acids in protein foods varies considerably. Animal proteins have a higher biological value than vegetable proteins. From an economic point of view animal and vegetable proteins are both used in practical poultry rations. The optimum percentage of total protein of suitable quality included in the diet depends on the class of stock, the purpose for which the stock is being kept and the daily feed consumption.

Fats and Oils

Fats provide the fowl with its second source of energy and can be made from carbohydrate. They can be of animal or vegetable sources, but to whichever class they belong all must be broken down before being absorbed and reconstituted in the fowl's body. High levels of fats sometimes cause digestive upsets and interfere with utilisation of other food nutrients, such as calcium. A reserve of energy fat is stored in the fowl's body, notably in the abdomen and as subcutaneous fat.

Fatty Acids

In addition to a requirement for fat the bird also needs certain fatty acids in the diet for health and productive reasons. Linoleic acid and oleic acid are the two most important ones and are naturally provided in the birds diet by vegetable oils. Animal fats are low in these two fatty acids.

Vitamins

Vitamins are required in minute quantities compared with the other food nutrients. They are essential to life, the actual amount for a particular vitamin depending upon environmental conditions, the diet and rate of growth or egg production. Some vitamins are synthesised by the bird itself, but many are not. An inbalance of vitamins may lead to serious disorders.

Vitamin A

Vitamin A can be provided by either pure vitamin A or its precursor, carotene. Carotene is converted into vitamin A in the wall of the intestines. The vitamin is usually provided by dry A and D3 preparations. Natural sources are grass meal and yellow maize meal.

Lack of vitamin A reduces growth rate. Mortality is high and resistance to disease low; salivary and tear glands may cease to function. In adults symptoms are less obvious, although egg production and hatchability may be reduced.

Vitamin D

On exposure to sunlight the fowl's skin can manufacture vitamin D. Of the natural foodstuffs fish oils alone supply any vitamin D worth mentioning. It is usually supplied by addition of dry preparations.

Lack of vitamin D causes rickets in chicks. Bones and beak become rubbery, and 'beads' develop at the ends of the bird's ribs. In adults hatchability is seriously reduced, and egg production may cease altogether. Eggshell quality is invariably poor. The relationship between vitamin D, calcium and phosphorus is a critical one. Each substance must be included in the ration in its correct amount and in relation to the others.

Vitamin E

Vitamin E is widely distributed in natural feedingstuffs, and it is unlikely for deficiencies to occur. When practical deficiencies are suspected it is not unusual to find some interfering substances, such as rancid oil, destroying the vitamin. Overcrowding, lack of feed trough space and disease are all examples of stress factors which may reduce the intake of vitamin E to the danger level. Chicks fed deficient diets show muscular inco-ordination. The condition is called *nutritional encephalomalacia* or *crazy chick disease.* The use of an antioxidant and the trace element selenium are related to vitamin E and spare its requirement.

Vitamin K

This vitamin is termed the *anti-haemorrhagic factor*. A lack of it results in haemorrhages on legs, breast and wings, accompanied by a delayed blood clotting time. Vitamin K is found in grass meal but is usually added as a synthetic preparation, menadione sodium bisulphite.

Vitamin B Complex

Members of this complex are thiamin (B1), riboflavin (B2), pyridoxine (B6), pantothenic acid, nicotinic acid, choline, vitamin B12, biotin and folic acid. All are essential in small amounts for optimum growth rate and production. The actual requirement for any member of the group varies considerably. Diets containing many feed ingredients of both animal and vegetable origin will furnish some of the bird's requirements. In modern high-nutrient density rations supplementation is always necessary to cover the requirement for fast growth rate and high egg production.

Ingredients such as fish meal, dried distillers solubles and dried skim milk are rich in B-group vitamins. It is usual to add synthetic sources of each to practical rations.

Minerals

The various minerals play an important part in avian nutrition. They are essential for the maintenance of body processes. The bird's skeleton contains most of the calcium and phosphorus, and potassium is found mainly in the muscles, iron in the blood and iodine in the thyroid gland.

Mineral requirements vary with the age of the bird and its sex. Many of the minerals are necessary in the diet of the fowl, but with few exceptions most can be supplied by the natural diet. Many, such as salt, calcium, phosphorus, iodine, manganese, zinc and selenium should be included as supplements to the natural ingredients.

Calcium and Phosphorus

The skeleton contains about 99% of the calcium and about 80% of the phosphorus in the body, and because of this, and because the ratio between the two minerals is critical, the two are considered together. Foods rich in calcium are meat and bone, fish meal, bone flour and limestone flour. With exception of limestone flour these foods are also rich in phosphorus. The metabolism of both these minerals is influenced by the vitamin D content of the diet and to a lesser extent by the supplies of manganese and zinc.

A good food store should be vermin proof and have a clean floor

Chick rations must contain all the calcium and phosphorus necessary for good growth, while laying hens may have supplementary calcium in the form of oyster shell grit. In this respect, sufficient calcium may be included to cater for a production level of 80% to 100%.

Inadequate levels of calcium and phosphorus lead to poor egg-shell quality and a form of cage layer fatigue caused by withdrawal of skeletal calcium. On calcium-deficient diets the laying bird can withdraw its skeletal calcium for the production of 5 to 6 eggs. High levels of fat in the diet appear to prevent absorption of calcium, while extremely low levels appear to encourage absorption.

Manganese

Lack of this mineral causes perosis and reduced hatchability. Egg-shell strength also decreases. Manganese is closely linked with calcium metabolism in that high calcium levels increase the manganese requirement.

Turkeys, ducks and geese also require manganese in amounts similar to those for the chicken, the actual amount varying according to the rations' energy level.

Salt

Salt improves the utilisation of dietary protein, and all the birds should receive sufficient (about 0.5% of the total ration). Excess causes high water consumption and death. A deficiency of salt causes hypertension in young and adult fowl and a cessation of egg production.

Iodine

A deficiency of this mineral will cause a reduction in metabolic rate resulting in lethargy. Iodine is usually incorporated into the mineral supplement included in the rations.

Zinc

Lack of zinc reduces body weight gain in young table chickens and turkeys, and the birds often adopt a 'goose-stepping' walk. Feather development is also poor. The bird's feet may also be affected by dermatitis; hatchability is also reduced. Modern rations should have supplementary zinc.

ANTIBIOTICS

The word antibiotic means 'against life'; that is, against the life of micro-organisms. Antibiotics are soluble, organic substances produced by the growth of micro-organisms. Many thousands of antibiotics are known, but only a relatively few are used as feed supplements. Penicillin, streptomycin, terramycin and aureomycin are the main ones in use today in this country but their use is restricted to disease control and therefore authority for them rests with the veterinary profession.

The degree of growth stimulation depends upon the environment. In new buildings, not previously occupied by stock, and in laboratory germ-free conditions the increase in growth is slight. Lack of response may occur when stock have been subjected to them in the same buildings for several years. The greatest response appears to occur when the environment is sub-optimal. The main improvements likely with an antibiotic supplemented food are:

(a) An increase in growth during the first 6 weeks of between 3 – 10%

(b) A saving in food conversion rate of 2½ – 5%

(c) A possible reduction in the mortality and cull rate

Feed antibiotics, are, however, unlikely to affect mortality rates caused by infectious diseases, e.g. salmonellosis and coccidiosis.

In reality it is likely that an antibiotic does not stimulate growth but allows more normal growth rate to occur. This may explain why no growth response is noticeable in germ-free conditions. Following the administration of an antibiotic in the food under normal environmental conditions the intestinal wall becomes thinner, with the result that more food nutrients are absorbed. Bacteria or germs which inhabit the intestinal wall are reduced in number, leading, not only to increased food absorption, but also to a greater appetite.

Some germs become tolerant or resistant to an antibiotic. When this occurs the response in terms of growth rate will be reduced. The phenomenon of resistant organisms is becoming increasingly common and research to find 'better' antiobiotics is continuously being carried out.

Generally, antibiotics are included in the food at two levels. To obtain increased growth promotion in fattening poultry a low level or nutritional level is used. This is usually less than 10 grams per ton of food. In contrast with this a 'high level' of 50 − 100 grams may be used for stress control and egg promotion. When a veterinary prescription is obtained higher levels are permitted for specific disease control.

The term *stress,* which briefly defined means 'any deviation from the normal', is often combatted by the use of antibiotics. Some conditions of stress are moulting, vaccinating, handling, debeaking and sudden food changes. Point-of-lay pullets often undergo severe stress, as the changes mentioned above occur at a time when the pullet's body is in the adolescent stage.

Unless exceptionally high levels are used in the food fed to ducks and geese very little benefit appears to be derived from using an antibiotic with these species.

HIGH−NUTRIENT DENSITY RATIONS

With modern hybrids, used in the production both of table poultry and eggs, the use of so-called high-nutrient density rations is now extremely popular. Rations high in energy and nutrients are more efficient in the conversion of food into meat or eggs than those low in energy and nutrients. Owing to the differences of opinion on how high high-nutrient is, it is difficult to define the term. In broiler nutrition it is usual to call broiler rations high-nutrient when they contain approximately 1400 – 1450 kcals M.E./lb (12.9 – 13.3 MJ/kg) and high levels of other nutrients.

For laying-hen nutrition the figures for guidance are lower than those given above for broilers. High-nutrient density rations result in less feed being eaten for each kg of meat or dozen eggs produced. It is important to remember that high nutrient density rations have higher levels of all other nutrients compared with rations lower in energy. The bird eats primarily to satisfy its energy requirement; thus the required level of nutrients must be concentrated according to the energy level of the rations.

With small laying hybrids the use of high-nutrient density rations is throughly recommended, for, with low-nutrient density rations, there is a risk of under-nourishing the high-producing layer. Heavy types of hybrids do not benefit by feeding high-nutrient density rations to the same extent. Their appetites are usually greater, resulting in a possible over-consumption of energy and other nutrients to the detriment of the bird's bodily condition and egg production.

Calorie/Protein Ratios

Protein and amino acids are used most efficiently when the ratio of energy to protein is correct. The actual ratio varies with species, age and breed of bird and widens with age; thus the younger the stock the narrower the calorie/protein ratio. The calorie/protein ratio means that for each 1% crude protein in the ration a certain number of calories are required. For example, the ratio for broilers up to 4 weeks of age is approximately 1% protein to 60 calories (M.E.) and from 4 weeks to marketing it is approximately 1 : 70. With turkey poults the ratio is narrower but widens appreciably, depending on the age at marketing.

With laying hens the ratios are more difficult to establish. When production energy and crude protein are again adopted as the criteria, the calorie/protein ratio is approximately 1 : 75. It cannot be stressed too much that the ratios given should be used only as a guide. They are not exact.

'PHASE FEEDING'

The term *phase feeding* indicates the use of one or more rations during the productive life of a bird. The system is, of course, applied widely in table poultry production but has not, until recently, been used in feeding for egg production. The theoretical reasoning behind the proposed system is as follows. At the commencement of lay the laying bird is still growing and, in fact, will do so until 9 − 10 months of age. Egg production is also increasing, as also is egg weight. The pullet's requirement at this period of production is greater compared with its needs later during the laying season. After peak production physical maturity is quickly reached, and any gain in body weight is almost certainly due to body fat deposition.

It is reasoned that feeding a high-nutrient density laying ration until physical maturity is reached ensures that peak production is obtained. A gradual change is then made to a ration lower in nutrients after this period with the result that body fat deposition is partly eliminated, production maintained at a higher level and a lower food cost per 12 eggs ultimately obtained. The lower feeding costs are brought about by a reduction in the cost per ton of food, after the changeover at physical maturity. The system of phase feeding is still new, and before being widely advocated it is necessary to know more about strain reaction and the effects of different climatic conditions on performance.

PRACTICAL FEEDING

Methods of feeding

The systems of feeding widely practised are dry mash *ad lib,* pellets *ad lib,* crumb feeding and mash and grain.

Dry-mash feeding is the most popular method for the battery cage system of management. Dry mash can be used for baby chicks, growers and adult poultry. Because the fowl consumes dry mash at a relatively slow rate the method deters cannibalistic tendencies. It is simple to operate and labour saving, as bulk-type food hoppers can be effectively used. Dry mashes should be coarsely ground using 3 − 4 mm sieve. Fig. 4.2 shows an outdoor feeder suitable for mash.

Pellet and crumb feeding is becoming increasingly popular in intensive systems of management. Both have the advantage over mash feeding in that, when fed in batteries, they do not foul the water troughs. They are less dusty to handle and flow easily in bulk food hoppers. Pellets and crumbs contain the same ingredients as mash and can be fed as a scratch feed to deep-litter birds. As the birds are able to get their fill far more quickly than with mash there is a danger of cannibalism and feather picking. Debeaking or low intensity lighting is therefore advised for pellet and crumb feeding.

The mash and grain method is advised only with special grain balancer mashes. The system is not advised for the specialist poultry keeper, as inbalances in the fowl's diet are not out of the question. Grain balancer rations should be used only by general farmers with surplus cereals available. The normal recommendations are for 50% grain and 50% balancer mash but other ratios can be used.

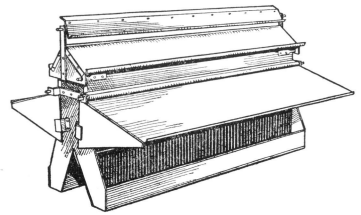

Fig. 4.2. An outdoor feeding trough for use on range with dry mash. The lids on either side protect the mash from the rain, and can be closed if required at any time to prevent the stock from having access to the food.

Soluble and Insoluble Grit

With laying rations containing a high level of calcium it is not necessary to provide soluble grit. When soluble grits are advised the feedstuff manufacturer's recommendations should be carefully followed. In the absence of these instructions the normal rate of supplementation is 3 — 4kg of oyster shell grit or limestone grit per 100 birds per week. With battery birds the shell should be sprinkled on top of the food.

Calcium grits should on no account be used to supplement chick, grower or turkey rations unless specifically advised.

Insoluble grit, such as flint or granite, is provided to aid the bird utilise its food. The combined action of the grit and the bird's muscular gizzard grind the food into small particles, thus increasing the surface area. This allows for greater action of the digestive juices and improved feed efficiency.

Insoluble grit can be purchased in varying sizes from chick size to large-turkey size. It is important to provide the correct size, for feeding too small a size may cause an irritation to the intestine and result in enteritis. Baby chicks should receive only token amounts of a chick-size, insoluble grit from 4 days of age. This should be sprinkled on top of the food. The change to a grower size should be made at 6 — 7 weeks of age and continued to 14 weeks. At this age a large or adult size should

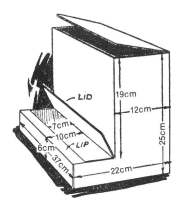

Fig. 4.3. Grit containers of the size seen here are large enough for any number up to 50 birds. In some a dividing partition is fitted to keep flint separated from oyster shell. Consumption may be controlled by a hinged lid. The more inquisitive members of the flock sometimes adopt the annoying habit of wasting grit by throwing quantities on the floor of the house. Where this occurs thin steel bars can be inserted, two inches apart and from front to rear of the feeding space.

be introduced. Turkeys should follow the same programme but be changed on to a large turkey size by 12 — 14 weeks of age. Under management systems involving deep litter the insoluble grit needs to be provided once each week. In systems other than with litter the feeding rate needs to be only once each month. A suitable grit hopper is shown in Fig. 4.3.

FEED REQUIREMENTS OF THE FOWL

The diets fed to poultry vary according to species, age and purpose for which the stock are kept.

The Chick

The chick grows extremely rapidly, and its nutritional requirements are extremely high, especially during the early stages. It is important to get chicks off to a good start. This necessitates the use of a high-nutrient density chick ration from day-old to 6 or 8 weeks of age. The protein content should be about 19% and the energy in terms of metabolizable energy 1250 – 1300 calories per lb of food. (11.5 – 12.0 MJ/kg).

The following is a typical ration for feeding to chicks during the first 6 weeks of life; no grain or calcium grits should be fed with this ration.

Chick Food	*Percentage*
Maize meal	35
Barley meal	10
Wheat meal	30
Grass meal	5
Fish meal	8
Soya bean meal	8
D.D.S. (Distillers Dried Solubles)	1¼
Limestone flour	1
D.C.P. (Di-Calcium Phosphate)	½
Synthetic vitamins*	¾
Salt	½
	100

*Vitamin A	8,000,000 international units per ton
Vitamin D3	2,000,000 international units per ton
Vitamin E	8 g per ton and full range of B-group
Vitamin K	2 g per ton
Minerals	80 g Manganese
	60 g Zinc
	10 g Copper
	20 g Iron

The Grower

Growers rations may be introduced after 6 weeks of age and fed to point of lay. The growing replacement pullet has a relatively low requirement for protein. Levels of 13% have been used with satisfaction although 14 – 15% is probably safer. The energy requirement will depend upon the environment. The energy requirement varies between 1200 – 1240 M.E. kcals/lb (11.0 – 11.4 MJ/kg).

The following ration is for intensively housed pullets, either floor or cage reared.

Growers food	Percentage
Maize meal	21
Barley meal	35½
Wheat meal	25
Grass meal	2½
Fish meal	3
Soya bean meal	8
Limestone flour	2
D.C.P.	2
Synthetic vitamins*	½
Salt	½
	100

*Vitamin A	6,000,000 international units per ton
Vitamin D3	2,000,000 international units per ton
Vitamin E	4 g per ton and full range of B-group vitamins
Vitamin K	2 g/ton
Minerals	80 g Manganese
	60 g Zinc

The Layer

The laying ration should be introduced at point of lay or 2 weeks before and fed throughout the laying season. Rations for light-type hybrids should be high in energy and protein. For heavier birds energy and protein requirements are similar, but because the birds have larger appetites the dietary levels are lower. The protein requirement is approximately 17% and the energy requirement will vary from 1230 — 1300 kcals/lb M.E. (11.3 — 12.0 MJ/kg).

The ration below is suitable for feeding to light hybrid stock. Supplementary oyster shell or limestone grit may be provided with the ration in very small amounts at the end of the laying period.

Laying food	*Percentage*
Maize meal	40
Wheat meal	30¼
Grass meal	5
Fish meal	6½
Soya bean meal	10
Limestone flour (or granules)	6¼
D.C.P.	1
Synthetic vitamins*	½
Salt	½
	100

*Vitamin A	6,000,000 international units per ton
Vitamin D	3,000,000 international units per ton
Vitamin E	4 g per ton and full range of B-group
Vitamin K	2 g per ton
Minerals	80 g Manganese
	60 g Zinc
	10 g Copper
	20 g Iron

The Breeding Hen

The breeding hen has similar nutritional requirements to the laying hen except for vitamins and minerals, which should be included in the ration at higher levels. In particular, members of the B-complex are very important, and the levels of vitamins A and D3 should be higher.

The level of animal protein is similar. The energy requirement is essentially no different from the laying hen.

Breeders food	Percentage
Maize meal	30
Ground oats	10
Wheat meal	$30\frac{1}{4}$
Grass meal	5
Fish meal	7
Soya bean meal	10
D.D.S.	2½
Limestone flour	3
D.C.P.	1
Synthetic vitamins*	¾
Salt	½
	100

*Vitamin A	10,000,000 international units per ton
Vitamin D3	3,000,000 international units per ton
Vitamin E	10 g per ton
Vitamin K	2 g per ton
Minerals	80 g Manganese
	60 g Zinc
	0.1 g Selenium

In addition B. Vitamins, Riboflavin, B12, B6, B1, Choline, Biotin Niacin and Pantothenic are essential.

Rations for Turkey

Some recommended formulae for turkeys are given in Table 4.1.

Table 4.1

Constituents	Starter	Rearer	Fattener	Breeder
Maize meal	30	20¾	30	20
Wheat meal	16¼	27	25	39
Barley meal	10	20	25¾	10
Fish meal	14	8	4	7
Soya bean meal	20	12	8	11
Grass meal	5	5	–	5
D.D.S.	2½	2½	–	3½
Synthetic vitamins*	1	¾	¾	1
Limestone flour	–	1½	3	2
D.C.P.	1	2	3	1
Salt	¼	½	½	½
	100	100	100	100

* The vitamin and trace element requirements are very high and should be met through the provision of a special proprietary turkey supplement.

It is advisable to add both an antiblackhead drug and coccidiostat to the starter ration when fed for the first 5 − 6 weeks of age. An antiblackhead drug should be added to both rearer and fattener rations to prevent blackhead from occurring.

The starter feed contains approximately 27% crude protein and should be fed from day-old to 5 weeks of age. No grain must be fed with this ration. The rearer food contains approximately 23% protein and is fed from 5 − 10 weeks of age. The fattener ration can be fed from 10 weeks until the turkeys are filled. The protein content is reduced to 15% and the calorie/protein ratio is widened to encourage a good 'bloom' or 'finish' on the marketed turkey carcass.

The breeder ration must be introduced 4 − 6 weeks before hatching eggs are collected. It is a complete ration, and only ½ oz grain should be fed to each breeding bird per day. It contains approximately 18%

Table 4.2

Average analysis of commonly used raw materials

	oil %	crude protein %	fibre %	ME Cals/lb %	ME MJ/kg %	Ca %	P %	Salt %	lysine %	methionine %	cystin %
							MINERALS			*AMINO ACIDS*	
Barley	1.5	10	4.5	1240	11.4	0.05	0.35	0.10	0.10	0.19	0.2
Maize	3.6	9	2.5	1540	14.2	0.45	0.24	0.10	0.3	0.2	0.18
Maize germ exp.	9.2	11	3.4	1590	14.7	0.15	0.5	0.10	0.58	0.19	0.2
Rice bran	14.0	14	6.15	850	7.8	0.75	1.4	–	–	–	–
Milo	2.5	9.6	2.4	1460	13.4	0.12	0.23	0.10	0.24	0.16	0.14
Wheat	1.5	10.5	2.5	1430	13.2	0.14	0.26	0.10	0.38	0.2	0.3
Wheatfeed	4.3	15.3	5.8	1180	10.9	0.24	0.68	0.10	0.7	0.3	0.35
Oats	4.4	10.2	10.0	1115	10.3	0.22	0.4	0.10	0.4	0.22	0.20
Cassava	1.5	1.7	4.0	1320	12.2	0.30	0.36	0.11	0.45	0.21	0.13
Potato flour	0.7	8.0	1.9	1495	13.8	0.07	0.18	–	0.40	0.18	0.11
Beans (field)	1.4	24	7.6	1110	10.20	0.13	0.39	0.10	1.65	0.22	0.09
Fishmeal Peru	9.0	65	0.50	1420	13.1	3.8	2.43	2.85	5.00	1.78	0.81
Fishmeal S.A.	6.5	67	0.50	1340	12.3	4.0	2.7	1.50	5.00	1.80	0.80
Fishmeal white	6.0	66	0.50	1250	11.5	7.5	3.5	1.50	5.00	1.80	0.80
Herring meal	9.0	70	0.50	1490	13.8	2.4	1.9	1.50	5.50	2.10	1.00
Meatmeal 55	4.8	55	2.0	790	7.2	8.3	3.90	1.30	3.10	0.60	0.50
Meat and bone 50	3.2	60	2.6	740	6.8	10.0	4.7	1.25	2.7	0.57	0.41
Meat and bone 45	2.1	45	2.7	700	6.4	13.0	6.0	1.60	2.1	0.50	0.40
Poultry protein meal	13	62	2.5	1200	11.0	3.3	1.7	0.35	2.7	1.3	0.98
Biscuit meal	18	8.0	4.5	1700	15.6	0.44	0.20	0.7	0.3	0.2	0.18
Grassmeal 15	2.5	15	22	490	4.5	0.95	0.29	1.31	0.7	0.25	0.12
Grassmeal 18	3.1	18	18	540	5.0	0.86	0.32	1.30	0.74	0.20	0.10
Feathermeal	2.5	85	1.5	1120	10.3	0.20	0.76	0.3	1.55	0.50	3.00
Dried poultry manure	2.1	11.0	8.3	450	4.1	7.00	2.3	0.45	0.39	0.12	0.21
Bloodmeal	1.0	80	1.0	1300	12.00	0.30	0.23	0.36	7.00	1.00	1.40
D.C.P.	–	–	–	–	–	25.7	17.40	–	–	–	–
Steambone meal	4	6.0	–	–	–	32.14	13.91	–	–	–	–
LSF	–	–	–	–	–	37.0	–	–	–	–	–
Salt	–	–	–	–	–	–	–	1.00	–	–	–

crude protein and is sufficiently high to maintain production over a 4 – month breeding season at low feed intakes.

Rations for Table Duckling

Table duckling can be produced satisfactorily by using a baby chick food for the first 3 weeks. From 3 weeks to killing age (between 7 – 8 weeks) a fattening food may be fed. For satisfactory weight gains the food must be fed *ad lib,* preferably as a pelleted feed. Special duck rations are produced by a few feeding-stuff manufacturers. When available they should replace the poultry rations mentioned. Breeding ducks may be fed on poultry breeders food supplemented with oyster shell grit. It is usually unnecessary and uneconomic to include either a coccidiostat or growth promotors in duck rations.

Rations for Guinea Fowl and Quail

Guinea fowl can very successfully be raised on rations fed to broiler chickens. However, certain coccidiostats are toxic to guinea fowl and clearance for use should first be obtained from a veterinary surgeon. Feeding is *ad lib* in crumb and pellet form for starter and finisher feeds, respectively. Breeding birds must be given a high quality turkey breeder ration of 17 – 18% protein, in order to obtain good hatchability. Quail for the table need a turkey starter ration of 26 – 28% protein. No other ration is needed. Breeding birds need a turkey breeder diet and both fattening and breeding birds should be fed *ad lib.*

BORROWER: **Anne-Mullane, Livestock/Poultry Secti**

No: **23279**   Date due for return: **21/5/2002**

TITLE: **Practical poultry keeping.**

I have received the above publication and accept respons
time by by
                    HAND or REGISTERED POST
         (If returning it by van please ensure it is signed fo

Signed: *Ann Mullane*   Date:

RETURN NOTED:
Received the above:

    Date:

                                                        Li

Loans Issue-receipt

2C, 2601

ity for returning it in

the Library)

_____

_____

rian

CHAPTER 5

HOUSING AND EQUIPMENT

INTRODUCTION

Houses and appliances represent a considerable proportion of the capital investment in the enterprises, particularly when the poultry farm is run on intensive lines. Much thought is therefore necessary on the selection of houses and their construction, because initial mistakes are easily made and costly to rectify.

Floors are usually constructed of wood or concrete. Concrete floors are to be preferred as they are permanent, vermin-proof, easier to clean and require no upkeep. They cannot be moved, however, and therefore cannot be used in portable buildings. Wood floors should be laid on brick piers or wooden piles to prevent dampness reaching them. Wooden piles must first be steeped in creosote before being driven into the ground. The actual wooden floor should be supported by bearers and anchored by bolts.

There are three types of roof: (a) lean type, (b) span roof and (c) uneven span. The lean type is easy to construct and most suitable for small buildings, such as fold brooders, store sheds and domestic poultry keepers. In large lean-to houses natural ventilation is not easy to regulate, and extractor fans may need to be installed. The span roof has a ridge which is an equal distance from each wall. Natural ventilation is not difficult, as outlets can be incorporated in the ridge. In wide-span broiler houses the walls may only be 1½ metres high. The uneven-span roof has one side a greater distance from the walls than the other. It is most useful for long, narrow-type buildings. Common roofing materials are asbestos, wood and roofing felt, corrugated iron and tiles.

Doors

Doors may be of two types: hanging or hinged type. Hanging types are preferable, as they are unaffected by wind and, if need be, they can be partly left open to supply extra ventilation in hot weather. A further advantage of the hanging type is that it is space saving. This is particularly useful in laying houses. Hanging doors are, however, more expensive to make than hinged ones.

Ventilation

The main object of efficient ventilation is to ensure an adequate and regular supply of fresh air to the birds, removal of unwanted gases and excess moisture vapour. The relationship of temperature to ventilation is also important. Poultry houses are either ventilated naturally or by mechanical means.

Natural ventilation

The efficiency with which the system works depends on convection and wind pressure to maintain air throughput. Warm air rises and if openings are made in the roof to allow warm air out and openings are made low down in the walls to admit cooler air, a throughput of air (ventilation) is obtained. Just how much ventilation is achieved depends on the size of openings, outside and inside temperature, and position of the wall openings. Wind speed is important because it can raise or lower ventilation rates.

The easiest way to provide a roof opening is to raise the complete ridge cap by some 5 cm. This permits the warm air to escape. For maximum effect the outlet in natural ventilated houses should be as far away as possible from the inlet area. So called ventilation cowls, either static or revolving type, can be used as air extractors. A cowl is a type of ventilation chimney which pulls used air out of the building. Air inlet areas should be equal to the extraction area. Windows can be used as inlets but such practise is best used in houses less than 8 m wide.

Natural ventilation is not recommended for intensive egg or meat production because the number of birds housed per square metre is so great that the open/closed window/cowl system would be totally inadequate. In most modern poultry houses today it is most desirable to exclude natural daylight from the building. This cannot of course be done with windowed houses! A further hazard is the lack of control of ventilation rate — a serious hazard indeed.

Mechanical ventilation

Electrically operated fans are used to alter the throughput of air. Most commonly, fans are located in the roof apex with the inlet ventilation points on the side walls. Fans may also be mounted in side walls with one set pulling in and the other set pulling out. This system is influenced by strong winds blowing against the building's side – result – uneven ventilation.

Pressure system

Fans pushing air into the building cause the air to be pressurised. The fan may be in the roof or side wall. Air under pressure can be made to reach all parts of the house and the usual method is via metal, hardboard and calico ducting. Another method is to bring air into the roof cavity and force it into the house through a fibre glass ceiling. With a pressure system the internal house surface must be moisture protected. The use of a vapour seal in the construction phase is advocated.

The number and size of the fans are chosen on the basis of the maximum air needed and also the minimum need in winter, for example, with layers the minimum ventilation requirement in summer is ⅓ rd to ⅙ th of the maximum.

Requirement for ventilation

Laying birds

As the adult weight varies only marginally it is only the minimum ventilation rate which needs to be selected.

A 4 lb hen (1.8 kg) for example, needs a ventilation system giving 10 m³/hour/bird (6 c.f.m/bird). This rate can be reduced to 10% or 1 m³/hour/bird as temperature falls.

The following table shows the maximum and minimum ventilation needed for laying hens of varying body weight.

Weight		Max. Required		Min. Required	
kg	*lb*	*m³/hour/bird*	*c.f.m./bird*	*m³/hour/bird*	*c.f.m./bird*
1.8	4.0	10	6	1.0	0.65
2.0	4.4	12	7	1.3	0.75
2.5	5.5	14	8	1.5	0.90
3.0	6.6	14	8	1.7	1.0
3.5	8.0	15	9	2.0	1.2

A solid floor field house mounted on wheels

Eggs being collected directly from battery cages into Keyes trays

Broilers and turkeys

In broiler and brooding houses heating is necessary and the ventilation is closely related to the changing body weight as the bird ages. The most sensitive measure of ventilation need is feed consumption and as a guide the allowance is 10 m³/hour/kg feed used per day (3 c.f.m. lb feed used per day).

The following table sets out the ventilation needed in a 10,000 bird broiler house at day old, 3 weeks, 5 weeks and 8 weeks of age.

Age	*Weight*		*Max. Ventilation*		*Min. Ventilation*		*No. fans*	*House*		*Fan*
(weeks)	*kg.*	*lb.*	*m³/h/brd*	*c.f.m/brd.*	*m³/h/brd.*	*c.f.m/brd*		*ºF*	*ºC*	*Spd %*
D.O.	0.05	0.10	–	–	0.1	0.06	0.1	85	29	0.10
3	0.60	1.4	–	–	0.7	0.40	0.7	70	21	0.70
5	1.10	2.50	–	–	0.85	0.50	0.9	70	21	100
8	2.0	4.40	10	6	1.30	0.80	1.5	70	21	100

Insulation

To get good air movement the house must be warm. Insulation attempts to achieve two things: first, it maintains house temperature during the winter and, secondly, it keeps the house cool during the summer. The value of insulating materials is judged by their effectiveness. This is expressed in Btu (British thermal units)/hour. The lower the figure the better is the material, and vice versa. Materials such as glass fibre, wood and insulation board have low U-values, whilst asbestos, glass and corrugated iron have high U-values.

The most important part of a house to insulate is the roof. Heat rises and, as the roof occupies over half the total surface area of the house, this is where most heat is lost. Poor insulation is not much better than none at all. Insulation material must be protected from dampness. In this respect, a good vapour seal is essential. Such materials are special paints, polythene and aluminium foil (also a good heat reflector).

In controlled-environment poultry houses the sides, roof and floor are insulated. The floor is usually insulated with hard core and roofing felt, while the sides are insulated with materials as used in the roof. Good insulation, although adding to the capital outlay, can save considerable money in the form of lower food consumption and better production. The additional cost per house may appear high but, when

A utility breed is the White Leghorn, this being in the light category

Modern controlled environment poultry house

calculated on a bird or square-foot basis, is negligible.

TYPES OF HOUSING

Turkey verandah

A verandah suitable for 35 fattening turkeys is shown in Fig. 5.1. The unit is 5.5 m long, 1.5 m wide and stands 0.6 m clear of the ground.

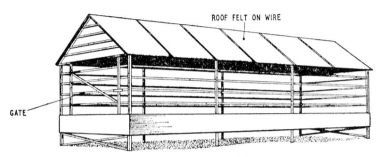

ROOF FELT ON WIRE

GATE

Fig. 5.1. Turkey verandah

The height from the floor to the ridge is 1.65 m. The floor is constructed of 3.8 cm x 1.5 cm wooden slats spaced 2.5 cm apart. Joists 6.6 cm x 2.5 cm spaced at 38 cm centres support the slatted floor. The main advantage of the turkey verandah is that the attendant can see the birds easily, and there is no need to enter the unit to water or feed the birds, as this equipment can be hung on the outside walls. Slats 6.3 cm apart allow the birds to put their heads through and reach food and water. The turkey verandah has been used for breeding in the past. Each unit holds ten females and one male.

Night ark

The night ark (Fig. 5.2), whilst primarily designed for rearing chickens on free range, can be also used as a fold unit when attached to a wire run. At 6 – 8 weeks of age the night ark can hold 40 – 50 growing birds. At 12 – 14 weeks of age the numbers should be reduced to 20 – 25 pullets. The ark has a slatted floor, usually divided into three sections for ease of removal and cleaning. Droppings trays may be provided but are unnecessary provided the ark is moved frequently to prevent fouling the ground. Ventilation is by means of an adjustable

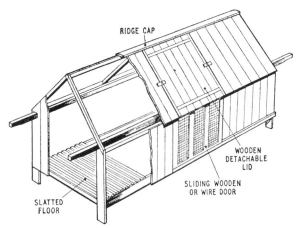

RIDGE CAP

WOODEN
DETACHABLE
LID

SLIDING WOODEN
OR WIRE DOOR

SLATTED
FLOOR

Fig. 5.2. Night ark showing slatted floor and general construction

slide in front of the unit and also by the raised ridge board. Framing is 5 cm x 5 cm timber and the unit measures 1.8 m x 1 m x 1.25 m to the ridge.

Nest boxes

The type of nest box shown in Fig. 5.3 is popular for egg production and breeding in floor management systems. Each box is 0.3 m x 0.35 m x 0.35 m high and because of its compactness offers the bird more

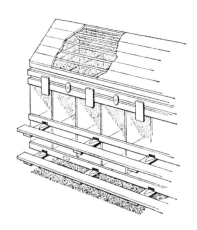

Fig. 5.3. Individual nests

privacy than the communal-type nesting box. Where several tiers of nests are used, alighting boards are essential to allow ease of access to the top tier. If necessary, alighting boards may be hinged and used to close up the nests at night. The roof should be sloped at an angle of 45⁰ to prevent the birds from using them for roosts. To increase privacy hessian sacking can be draped over the nest fronts. To discourage floor laying, which is particularly common in wire-floored houses, the nests must be made attractive and comfortable. They should be sited with their fronts facing the lowest light intensity in the house and should not be placed too high from floor level or the birds may refuse to use them.

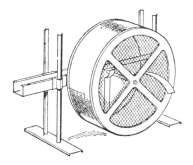

Fig. 5.4. Food cleaner. As the moving food in the troughing comes to the cleaner, it is ocnveyed inside. Any litter, etc., are removed by the cleaner and thrown out on to the floor.

The lowest tier should be 20 cm from the floor and the highest tier no more than 1 m high. Straw nest mats or rings may be used as a floor covering. Wire-floored nests should be covered with straw at the commencement of lay to encourage nest box usage.

There is no need to replenish the supply once nest box laying is established. So as not to discourage nest box usage, egg collection frequency should be substantially reduced at the start of lay. This will permit the nervous layer to explore the surroundings without disturbance which is often a cause of encouraging floor egg laying. 0.093m³ of nest box space should be allowed to every three or four pullets. The former figure is preferable for high-producing, egg-laying strains, particularly during peak production.

EQUIPMENT

Automatic feeders

There are many types of automatic feeders on the market, and it would be impossible to describe each in detail. Generally speaking, it is uneconomical to employ an automatic feeder with fewer than 5,000 adult birds, for although expense is reduced with regard to feeding space the driving mechanism and bulk-food drum is not reduced proportionately to make the investment economic. Chain or automatic feeders are generally less wasteful than the conventional tube-type of feeder (discussed later) and they also stimulate the bird's appetite by continuously moving the food. Four birds should be allowed for every 0.3 m of feeder space, and the trough circuiting must be so placed that no bird has to travel more than 2.5 – 3.0 m before reaching food. This is particularly important in wire-floor houses where stocking rates are extremely high. In houses where some littered area is used a litter 'cleaner' (Fig. 5.4) should be incorporated in the feeder to prevent blockages and breakdowns. The chains must be checked periodically to ensure that surplus links do not cause mechanical damage. This is usually done by a chain adjuster which can be fitted to every circuit. It is important that there is correct chain tension for smooth working. Some automatic feeders can be run horizontally or vertically and can pass over or under obstructions. They can often serve more than one building or even multi-storey units. A hopper supplying the feeder 2 m x 1 m x 1 m has a mash capacity of 275 kg and is capable of conveying 114 kg each hour.

One automatic feeder can be used for floor management systems or in batteries. It can also be used for filling tube feeders. The tube feeders are filled by gravity (see Fig. 5.5).

Watering equipment

The fowl's water space requirements should not be judged by the length of the troughs but by their location and spread. No bird should have to walk more than 3 m for water. As many birds have to travel more than this distance in intensive poultry houses the question of distance explains why battery caged layers usually lay better than those housed under other management systems. It is essential to allow one 2 m double-sided water trough to every 100 laying birds. The depth of water in the trough should be at least 2 cm and the drinker should be at least 5 cm wide. Circular drinkers are very satisfactory for all ages of

An automatic drinker for a deep litter house. The water level is controlled by a float-operated valve, and the detachable cage helps to keep the bowl clear and prevent flicking. It is used with mains or a static tank.

A shaft ventilator common to incubation houses. A gauze-covered opening at the base of the shaft and inlet to the house admits the fresh air. Current of air is controlled by the slide.

Houses should be creosoted regularly, and a pressure sprayer greatly assists such work, besides being useful for many other purposes.

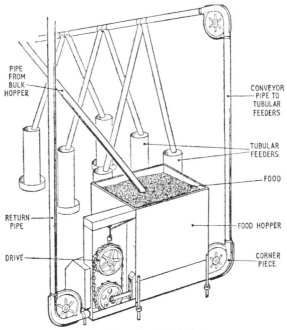

PIPE
FROM
BULK
HOPPER

CONVEYOR
PIPE TO
TUBULAR
FEEDERS

TUBULAR
FEEDERS

FOOD

RETURN
PIPE

FOOD HOPPER

DRIVE

CORNER
PIECE

Fig. 5.5. Automatic feeder

poultry. One 20 cm diameter drinker is sufficient for 100 birds. Drinkers should be adjustable so that cleaning and alteration to height may easily be carried out. With the circular type it is advisable to provide a simple drain or spillage pan. These are necessary in deep litter houses but unnecessary on wire – or slatted – floor units.

Nipple and cup drinkers

Nipple or valve drinkers are the most common type of drinker used in commercial egg production. The nipple consists of a valve which is set into an alkathene piping running the length of a battery cage flock. Water pressure keeps the valve shut. Action by the hen forces the valve into an upward and open position and water then runs into the birds beak. At least one nipple drinker is needed for every four laying birds. Equally important is the siting of the nipple. It must be easily seen, readily accessible and for preference each bird should have a choice of two nipples to drink from. Daily checking to see that the nipples are both working, yet not flooding, is important.

Water cups consist of small plastic cups situated usually on the outside of the cage. Each cup has a small amount of water, the level being governed by the weight of water. As with the nipple drinker regular inspection is necessary.

CHAPTER 6

MANAGEMENT OF LAYING STOCK

LIGHTING

The practice of providing artificial light to extend the natural daylength
has been practised by commercial egg producers for many years. Light
rays reach the bird's pituitary gland at the base of the brain via the eyes.
This results in the release of hormones which circulate in the blood
stream and stimulate ovarian activity. Extended lighting for 14 – 17
hours in each 24 hours causes increased egg production.

The use of controlled-environment laying houses necessitates complete reliance upon artificial lighting and ventilation. Under this system
of laying house management it is usual for the birds to start laying
when the total daylength is between 6 and 12 hours. The difference in
daylength is dependent upon the system of rearing and amount of light
provided when the birds are at point of lay. It is usual for the light to
be increased by 15 – 30 minutes per week once production has started
or is approaching. By the time peak production is reached the birds
should be receiving not less than 12 hours of total daylength. Nothing
appears to be gained by increasing the daylength beyond 17 hours. The
intensity of light required depends to a great extent upon the system of
management. In battery cages where the risks of cannibalism etc. is
high the intensity must be of the order of 0.05 lux. In wire-floor houses
where stocking densities are high the light intensity should be low. In
deep litter houses with high stocking rates the intensity should also be
about 0.05 lux. For lower stocking rates 0.3 lux is suitable.

In naturally-lighted houses a side-windowed area of 0.093m^2 of glass

The sheltered section of a straw yard, which contains a droppings pit, nesting room, and feed and water troughs

In this instance old farm buildings have been converted for use as a straw yard

for every 0.116 – 0.130 m² of floor space and roof-windowed area of 0.093 m² for every 0.26 m² are recommended. In windowless houses 60-watt bulbs may be placed 5 m apart and hung 2 m from the floor. Depending on the reflection from the internal surface, the light provided will be about 0.03 – 0.07 lux with 15 watt bulbs.

There are three main lighting systems which can be used in windowed poultry houses: evening lighting, morning lighting and a combination of both. With evening lighting and a system of housing incorporating deep litter and perches it is necessary to use a dimming device so that the birds are not plunged into darkness when the lights are switched off. With morning lighting this is unnecessary.

There appear to be no advantages of one system over the other in terms of annual egg production.

Calor gas and paraffin can be used as alternatives to electricity.

POULTRY MANURE

Poultry manure is an extremely valuable source of nitrogen, organic matter, phosphate and, to a lesser extent, potash. The exact value of the manure depends upon a number of factors, of which the following are the most important:

(a) Quality of moisture in the material
(b) Amount and nature of litter material used
(c) The nutrition of the birds from which it is obtained
(d) Storage of the manure

The drier the manure the more concentrated will be its plant food or fertilising ability. The most common types of litter material used are wood shavings, chopped straw, peat moss and sawdust. All of these make excellent basal structures for manure. Peat moss is the most absorbent, but its price makes it too uneconomic for use. Diets high in protein produce manure with a higher nitrogen content compared with lower protein rations. Hen manure is higher in nitrogen, phosphate and potash than duck manure. Poultry manure which is left standing in the open may quickly lose its value through 'leaching' by rain water. It should always be stored under cover in a dry building.

The value of good poultry manure in comparison with farmyard manure is shown on the following page.

	% Nitrogen	% Phosphate	%	Potash
Broiler Manure	2.8	3.0		1.5
Deep litter hen manure	2.4	2.5		1.5
Farmyard manure	0.6	0.3		0.5

The minimum value of poultry manure should be obtained by costing it on the basis of its nitrogen, phosphate and potash content. The unit cost of these nutrients should be based on the subsidised straight fertilisers, sulphate of ammonia, superphosphate and unsubsidised muriate of potash. Based on the value of certain organic fertilisers the broiler manure given above should be priced around £7.00 a ton.

Quantities produced

As a rough guide, poultry produces about 1 kg of fresh droppings for every ½ kg of food consumed. In the case of adult laying poultry 1,000 head produce about 95 tonnes of fresh droppings each year. In broiler production the amount of manure, including the litter, is about 2¼ tonnes for every 1,000 birds over a 10-week production cycle period.

METHODS OF APPLICATION TO FARM CROPS

The natural way to apply poultry manure is by folding poultry over grassland. With this system losses of nitrogen are avoided, and soluble constituents are retained in the soil. Portable poultry houses, fold units and hay-box units should be moved frequently to ensure an even distribution of the manure over grassland area.

Manure from broiler and deep-litter laying houses is most effectively spread mechanically at the rate of 5 tonnes to the hectare. Droppings from battery birds are more difficult to dispose of because of their higher water content in comparison with a mixture of manure and litter.

In large battery units it is not unusual for the droppings to be conveyed mechanically from the houses into a septic tank. The so-called 'slurry' which forms is sucked up into a mobile tank and sprayed over grassland. It can with local authority permission, after adequate fermentation, be channelled into the main sewage pipe lines for disposal. When estimating the tank capacity for holding 'slurry' allowance should be made for twice the volume of water. Installing a storage tank and the

*Interior view of small deep litter pen showing food and water troughs
and nest-boxes*

*Roosting end of a fold unit showing slatted floor, nests and broody coops on
top right. The lid of the nests opens upwards*

purchase of distribution equipment is probably not justified economically with fewer than 5,000 laying birds. The tank may be sited under the
floor at the end of the battery house.

A satisfactory underground tank at the end of a two-stack battery unit of 3,000 birds should measure 5 m x 3 m x 2 m. The walls of the
tank should be pre-cast blocks at least 15 cm thick. The capacity of such
a tank will be 320m², which will take the droppings from 3,000 battery
birds for 45 days. Cages with paper cleaning cannot be used in conjunction with this method as the paper will not 'break down' to pass through
distribution pumps and sprinklers.

In wire-floor and deep-pit laying houses the droppings can be
removed once a year by mechanical or manual labour. The droppings
under the wire floors should be in a crumbly state if the ventilation of the
house is correct. The addition of superphosphate at the rate of 25 kg per
100 birds per month scattered through the wire over the droppings will
make the manure dry, more friable and easier to distribute.

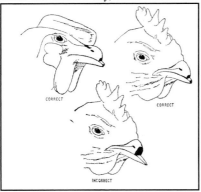

FIG 6.1   Debeaking
This avoids cannibalism
which occurs in intensive
systems. The ethical aspects
are against the method.

MANAGEMENT METHODS

The methods available are as follows:

1. Battery or Laying Cages, where birds are kept in confinement.

2. Free Range, where birds have to be given considerable open space,
generally no more than 100 birds per acre (EU law is 1,000 birds per
hectare = 2.471acres, so this is more generous). Usually laying birds are
in sheds containing, say, 100 birds which have access to a field where
they can forage. Food such as layers' pellets are inside so the birds can
return to feed and lay. The system requires considerable labour to feed
the birds, although it can be mechanized by siting the houses in a way
which allows them to be limked by automatic feeding sytems.

3. Semi-Intensive Systems:

(a) **Deep Litter**; (b) **Straw Yards** or **Hen Yards**; (c) **Aviary/ PercherySystems**; (d) **Fold Units**. These may be varied to incorporate, where appropriate, weld mesh flooring or slatted floors, which also applies to free range systems which may be in very large houses (1000 birds or more) or smaller units not exceeding 100 birds. The Aviary/ Perchery systems consist of a large house and run with perches and shelves to give plenty of exercise and are ideal for the light breeds.

Because of the higher costs of full free range and shortage of land the *semi-intensive systems* are likely to be adopted. They allow a greater concentration of birds, but management must ensure that hygiene is to a very high standard. In deep litter the litter such as shavings, short straw, or peat moss should be built up so it is quite deep, with the result that the bacteria breaks up the droppings quite automatically. It must be kept dry by adding more litter (to 25cm). The importance of the systems is that they give scratching areas and therefore the birds are kept busy. Overcrowding must be avoided and stocking density not more than 3 birds per square metre for straw yards and deep litter 7 birds m^2.

Eggs from the different systems must be in marked cartons with date.

Water trough with covered ball valve. ~ ~
Communal nests on floor at end next to egg packing room and communicating door. ~
Food troughs suspended in gangway. ~ ~ ~ ~ ~
Enclosed droppings pit occupies two-thirds of shed. Perches are 14 ins apart on 1½ ins mesh netting
Deep straw litter in shed & yard
Detachable front panels of glass and corrugated iron. ⟶

Litter kerb of old sleepers

Bruff Jackson

SEMI-INTENSIVE:

Hen Yard,

Straw Yard

or Barn System.

Goes under

different names.

Could be

adapted as an

Aviary or

Perchery System.

Birds can exercise

and get fresh air

and are quite

active.They do

not become bored

or develop **Battery**

Cage Syndromes.

Laying Cages

A laying battery consists of a block of individual cages, each one designed to hold 3 up to 40 or so birds. The cages have sloping wire floors through which the droppings fall, but eggs are retained. The droppings may fall and be collected on glass trays, boards, paper, asbestos or belts. Food is supplied in troughs attached to the cage and filled either by hand or by a travelling bulk hopper. Water is supplied in metal or polythene guttering above the food trough or by nipple valves and small cups.

Three-tier cages are most commonly used in this country. Four-tier cages may be used in buildings which have plenty of space to the eaves and ridge. The dimensions of an individual laying cage are 22 – 30 cm wide, 45 cm deep and 45 cm high at the front sloping to 35 cm high at the back. The floor extends about 15 cm beyond the cage front to form an egg collecting tray.

Cages which are designed for more than one bird usually have the same general dimensions as the individual cages, but the width increases according to the number of birds per cage.

The following are typical widths of multiple-bird cages.

Cage width in c.m.	*Number of birds*
25	Single bird: light (1½-1¾ kg) or medium sized (2 kg-2½ kg)
30	3 light or 2 medium sized
42	4 light or 3 medium sized
50	5 light or 4 medium sized
78	8 light or 7 medium sized
100	9 – 10 light or 7 – 8 medium sized

It is important to remember that the number of birds per cage is governed primarily by the amount of food and water trough space available. Food trough space should be 10 – 12 cm per bird and water trough space 6 – 8 cm per bird. The number of birds per cage and choice of cage will depend primarily upon the number of birds to be housed in the battery house. Table 6.1 shows the relationship between width of cage, number of birds, number of cages and floor area in houses varying in width about 18 m long. In this table a 1.5 m clearance space is allowed at each end of the battery blocks for food storage, manure removal or egg storage.

Table 6.1

Arrangement	Three birds in a wide cage			Seven birds in a wide cage		
	No. of Birds	No. of Cages	Floor space per bird in m²	No. of Birds	No. of Cages	Floor space per bird in m²
6 m wide Two blocks lengthways	1,440	480	0.070	1,596	228	0.060
8 m wide Three blocks lengthways	2,160	720	0.064	2,394	342	0.057
10 m wide Four blocks lengthways	2,880	960	0.061	3,192	456	0.055
8 m wide Eight blocks cross ways	1,812	604	0.078	2,016	288	0.065

The approximate height of a three-tier battery cage unit is 2 m and the width $1.2 - 1.4$ m. To prevent accumulation of rubbish under the lower tier it should clear the floor by 15 cm. This is also essential for a satisfactory ventilation system.

Four-tier units are some 2.5 m in height and $1.2 - 1.4$ m wide. The main disadvantage of the four-tier unit is the difficulty of servicing the uppermost tier. This is virtually impossible without the use of a movable trolley to stand on.

The simplest method of removing droppings from under the cages is by the use of a movable tray. This system is nowadays used only by the backyard poultry keeper. The two most popular systems of manure removal are mechanical scrapers or ploughs and the endless belt method. With the latter principle the belt passes from tier to tier and is scrubbed at each end when the droppings are scraped off into a pit or into hand trays (Fig. 6.2.) 'Cafeteria' laying batteries have their food and water troughs mounted on tracks or carriers which move backwards and forwards or around the unit in front of the birds. As the food and water trolley approaches the birds often become excited. This excitement can lead to vibration of the cage floors and broken or cracked eggs. The main disadvantage of the 'cafeteria' system is the risk of mechanical breakdown, which, if it occurs, leaves the birds with neither food nor water. The droppings also mount up. (Fig. 6.3).

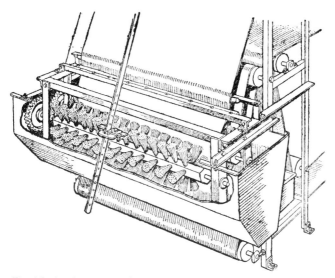

Fig. 6.2. A refinement in the cleaning of droppings in a laying battery.
The 'tray' consists of an endless belt which is not only brushed clean of manure but also passes through a bath of disinfectant.

The number of birds which may be managed by one man is usually limited by the egg collection. Most battery cage units have conventional floors into which the laid egg rolls; thus hand egg collecting and packing is necessary. A limited number of cage manufacturers offer battery cages fitted with automatic collecting devices. They usually work on an endless-belt system, the eggs rolling on to a tray attached to the movable food trolley. The eggs have to be collected from the bulk moving tray, but the labour in collection is considerably reduced.

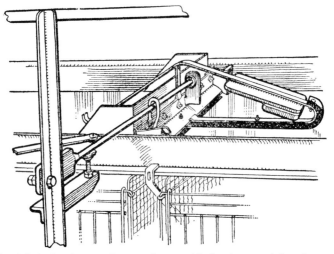

*Fig. 6.3. A plough type of scraper for use with glass bottomed droppings trays.
It moves in conjunction with the automatic feeding and watering trolley
travelling along the cages*

Colony cages

Colony or multiple-bird cages are battery cages which are constructed so that they are capable of holding 20 or so laying birds. The height and width of the colony cage is identical to the normal battery cage but differs in that it is strengthened both lengthways and sideways. This is essential because the total weight may not always be evenly distributed. Colony cages became popular in this country because the cage cost per bird is considerably reduced due to a high stocking rate caused by the absence of cage divisions. The normal back and dividing division is absent in the colony cage and whilst food troughs are provided on both sides of the unit only one central water trough is required. It is important however, that a wider than average water trough is provided so that birds on both sides of the unit can drink comfortably at the same time. Where nipple drinkers are used they should be sited on both sides of the cage and not down the centre.

The management of laying birds in colony cages is essentially the same as for birds in ordinary battery cages. Cleaning or manure removal may however need to be carried out more frequently because the stocking rate is higher.

Californian cages

A more recent development in this country is the introduction from the U.S.A. of the Californian cage. This differs considerably from the battery cage popularly used in this country. It has no dropping trays, and the tiers are not placed one above the other but are erected in pyramid fashion so that all the droppings fall clear of cages offset below the higher one. The main advantage of this type of unit is its comparatively low purchase price, ease of erection and saving in labour requirements. The droppings may be allowed to accumulate for at least 6 months below the cages before being disposed of. A disadvantage of the Californian cage is the additional space it occupies in comparison with the ordinary battery cage.

Interior of a deep litter house. The bundles of straw prevent birds from roosting on top of the nests

Deep pit

The deep pit system is in effect a battery cage system built over a deep pit so that the hen droppings can drop beneath the stock and remain there for up to 3 years. The pits are of varying depth from 3 m to 6 m below the birds. Deep pit houses should be ventilated by pulling the air through the roof and out beneath the birds but over the droppings. This system ensures that only fresh air reaches the stock and ammonia fumes are removed before they reach the birds. The warmed air also helps to dry the manure to a moisture level below that possible with a reversed method of ventilation.

Flat deck cages

Conventionally, cage blocks are either upright 3/4 tiers or A type stacking (pyramid type). However, cages may also be built on a single level with or without gangways between blocks.

The flat deck system may be built over a pit or incorporate traditional cleaning equipment. Bird performance is generally lower than in tiered cages but so much depends on husbandry practise that comparison of cage systems is difficult.

The wire-floor system

The wire-floor system of managing laying poultry was imported into this country about 15 years ago from the U.S.A. It enjoyed a certain amount of popularity, but during the last 10 years or so supporters of the system have declined in number, mainly in favour of the battery cage. The wire-floor system in its simplest form is a gigantic battery cage into which comparatively large numbers of laying birds are crowded. The normal stocking rate is one bird to 0.093m^2 of floor space. The floor is constructed of 7 cm x 2.5 cm welded mesh (usually 12-gauge) and is supported by bearers every 60 – 66 cm. It is recommended that perches are attached to the top of the mesh at 60 cm centres to ensure even perching distribution. Feeding and watering points should be plentiful so that no bird has to walk more than 3 m before reaching them. For preference, nest boxes should be placed along each side wall of the house to encourage the birds to use them and prevent floor laying. Siting the nest boxes on one wall or against the service passage side only makes it extremely difficult, and indeed often impossible, for birds occupying the opposite side of the building to reach them without much physical effort.

Results from the system are extremely variable, although with high standards of management they can be excellent. Errors in management and layout of equipment are quickly enlarged upon, and the wire-floor system, like that of the colony battery cage system, is dependent upon more than adquate food and water trough space, nest boxes and ventilation.

Slatted-floor system

The main difference between this system and the wire-floor system is the type of floor material used. The slatted-floor system involves the use of slatted wooden sections approximately 4 m long and 2 m wide, and each slat measure 2.5 cm across the top, tapering to 2 m at the bottom. The slats are spaced 2 m apart. There is no need, as with a wire-floor unit, for perches to be placed on top of the slats. Supporting wood framing and bearers need to be less abundant than with the wire-floor unit. Although it is necessary for the attendant of a wire-floor unit to confine his movements to the areas of greatest support this is usually unnecessary in a slatted-floor house because of the greater structural strength of the wooden slats. This system is not popular in this country because compared to battery cages, performance is inferior.

Deep litter system

Apart from the popularity of the battery cage system of management, deep-litter housing is still used by many of the smaller poultry keepers and, of course, by broiler breeders. The reason for this is not difficult to find, for existing buildings on general farms are often readily and easily adapted to the system. Further, when capital is limited, for a moderate outlay the small poultry keeper is able to commence egg production, whereas a battery unit not only requires the shell of the building but the cages also. The capital invested per bird can be similar — hence the tendency towards battery cage egg production.

The deep-litter system houses less birds per unit area than any other intensive system and yet, because of the necessity of keeping the litter dry and in working order, it requires good roof insulation and ventilation. The system offers no advantages over other management systems, and for this reason it will be discussed only with wire- and slatted-floor units utilising one-half or one-third littered-floor areas.

Wire or slatted-floor system with half-or third-deep litter
for breeding stock

The system incorporating wire or slatted floors and a littered area allows greater bird concentrations compared with the complete deep-litter system. Stocking rates are invariably increased by one third, as, in comparison with an allowance of 0.28 m per bird on deep litter, a recommendation of 0.18 m for part wire or slats and litter is made.

The wire-floor, part-litter system has been popular for the production of broiler breeding eggs, as the littered area allows the heavy type of bird used some measure of exercise. Some grain is fed to encourage exercise and keep the litter in working condition.

To help maintain the litter in a dry condition the water troughs should be situated over the wire-floor area. Where half wire and half litter is used 0.09 m^2 of each area is allotted per bird. Where one third wire is used, approximately 0.07 m^2 is allowed per bird and 0.14 m^2 per bird for the littered area.

Litter materials

The main considerations in choosing litter materials is their ability to absorb moisture, to decompose readily and their availability and cost. Litter should not be dusty, not tend to pack when wetted or be so light that it blows about the house. In addition to absorbing moisture the litter should dry out rapidly. The moisture capacity of some litter materials is given below:

Litter materials	*Weight of water absorbed per 45 kg of litter in kg*
Peat Moss	183
Oat Straw	127
Sawdust (oak)	45
Mixed Sawdust	68
Wheat Straw	117
Wood Shavings	154

Straw should be chopped into lengths of 15 − 20 cm, as it tends to matt down in the long state and takes a long time to decompose. Chaff is not recommended as a litter material because its absorbent abilities are poor. Wood shavings are the most popular material.

Verandah units

The verandah laying unit (Fig. 6.4) is an extremely economic and efficient method of commercial egg production for the small poultry keeper. The overall measurements of the verandah unit are 7.62 m by 2.6 m and approximately 1.67 m high. Depending on the size and weight of laying bird housed, each unit has a capacity of between 60 and 80 birds.

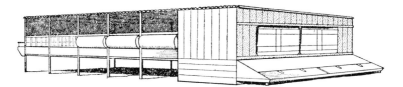

Fig. 6.4 Verandah laying unit. Suitable for 50 adult laying birds. Note the rollaway nest boxes. This type of unit is also suitable for rearing poultry

The floor is constructed of 7.5 cm x 2.5 cm 12-gauge welded mesh, on top of which are 5 cm x 3 cm wooden perches at 0.6 m centres. The sleeping and laying quarters are at one end, and in many verandah units a rollaway-type nest box is fitted. Food and water troughs are situated on the sides of the wire run in such a way that the attendant can re-plenish them from the outside of the unit.

Droppings fall through the mesh and may be allowed to accumulate until the end of the laying period, when the verandah can be removed for overhaul and disinfection. Artificial light is simple to install and can be linked up to other verandahs if required. This type of unit may also be used for rearing poultry from 8 weeks of age.

Breeding has also been successfully carried out in verandah units but the system is uneconomic for commercial production.

The annual clean out

All intensive systems of poultry husbandry that involve high stocking rates are especially prone to the build-up of disease organisms. Such build-up effects a reduction in growth rate and egg production and an increase in the incidence of secondary invading organisms, which inevitably lead to lowered production efficiency. Reduced production efficiency, the sum effect of reduced growth or egg production and high mortality owing to the build-up of disease, is apparent on examination of the profit and loss account compared with that when the housing and equipment were new. New buildings and equipment invariably effect a greater output than buildings and stock that have been in operation for a number of years without being completely disinfected. This is because harmful micro-organisms, present where stock is intensively housed, quickly multiply. With periodic cleaning of the houses and occasional breaks of its throughput schedule, the multiplication rate is to a certain extent controlled, and mortality rates·may be prevented from noticeably increasing. However, the steady build-up of infection is sufficient to lower the bird's resistance to secondary invading organisms, which, as they multiply, affect the rate of growth by lowering the efficiency of feed utilisation. The number of culls also increases.

The annual clean out, by removal of the stock from the houses for a certain period, is intended to break the reproduction cycle of harmful

A 'cafeteria' laying battery. Trolleys carry food and water troughs along the cages

or pathogenic micro-organisms. The duration of the 'no-production' period will depend on available labour to carry out the disinfection. The longer the rest period the more effective the operation. Two weeks is usually considered economically adequate.

All portable equipment should be removed from the building before disinfection begins. Litter material and faeces should be dampened down with a disinfectant, removed and, if a disease has occurred, burned or buried. The equipment should be soaked in a 4% solution of washing soda applied through a high-pressure steam cleaner. The house should be thoroughly cleaned before putting back the equipment. All windows and doors may be left open to let in as much light, sun and air as possible.

Disinfectants used should have a low cost-per-unit value; they must be safe and simple to apply, effective against a wide range of organisms and stable when exposed to air and light.

Records

For success in modern poultry production a definite plan of work should be developed, entailing a budget of receipts and expenses. To do this efficiently a number of day-to-day records are essential. The information so obtained may appear unimportant in the first instance but will prove very valuable as it is accumulated and analysed.

Records should be easy to compile and simple to analyse. Keeping records involves not only recording food consumption, egg production and body weight gain, but also other valuable information so that future planning, organisation and expenditure is possible.

At the end of the year the following questions should be asked and answered:

(a) What was the profit per bird per house and per square metre?
(b) What were the items of greatest expense? Can they be reduced without sacrificing performance?
(c) What were the items of greatest receipts and can they be increased without incurring excessive cost?

These questions can only be accurately answered through the use of records and their careful analysis.

In commercial egg production daily records of egg production and feed consumption should be kept. These should be transferred to graph paper on a weekly basis so that the flock performance may readily be followed. At the end of the year total costs and receipts will show a new margin for the unit as a whole. From these figures can be worked out the food costs per 12 eggs produced and food conversion rate. Egg grading figures should be used not only to indicate the various grades but also to provide information on second-quality and cracked eggs. This latter class can, of course, be caused by bad egg handling on the farm, poor feeding, housing, disease and heredity.

To calculate the percentage production divide the week's production by seven, multiply this by 100 and divided by the number of birds housed; i.e.

$$\frac{\text{Week's production} \times 100}{7 \times \text{No. of birds housed}}$$

The food cost per 12 eggs is obtained by first finding the F.C.R. (kg of food eaten to produce 12 eggs) multiplied by the cost per kg of food purchased; e.g. if the F.C.R. is 2 and food cost per kg of food 12 p then the food cost per 12 eggs is 2 x 12 p = 24 p.

In the production of table poultry, turkeys and ducks the food accounts for about 70% of the total cost. Therefore the F.C.R. (kg of food eaten to produce kg of liveweight) is important. This multiplied by the cost of 1 kg of food will give the food cost per kg of liveweight gain; e.g. in broilers the F.C.R. is 2.00 : 1 and the food costs 15 p per kg. Therefore the food cost per kg of liveweight gain is 30 p. Grading results are also vitally important in table poultry production for poor grading at the packing station can be the result of bad handling, poor feeding, disease and the wrong distribution and positioning of water and food troughs.

Side View of Hen Yard or Barn System
Detailed operation on page 92

Chapter 7

CHICKEN BROILER AND CAPON PRODUCTION

INTRODUCTION

The term *broiler* is used to indicate a mass-produced table chicken of either sex weighing about 2 kg liveweight and aged between 6–8 weeks. The actual weight range varies from 1½ – 2½ kg.

Comparatively speaking the chicken-broiler industry is new to the United Kingdom but, without doubt, it is the most highly specialised form of poultry keeping. In this country the production of broiler meat is in the region of 340,000,000 birds per annum. On a *per capita* basis this is approximately 6 birds per person per year which, in comparison with figures from the United States, is low.

The industry is extremely specialised, requiring not only a good knowledge of mass-production techniques but also big-business know-how. Today a large number of broiler chickens are grown on a contract basis in which the operator is a paid manager drawing a fixed sum of money each month. The number of birds each rearer produces is governed by the packing station, who market the final product. It cannot be over-emphasised that for success in the broiler business a guaranteed outlet is essential.

STRAINS USED IN BROILER PRODUCTION

Nearly all the broilers produced in the United Kingdom are the result of American breeds and breeding techniques. Four main organisations are currently marketing broiler chicks, and their ancestry is based upon the following breeds: White Plymouth Rock, Cornish White and New

Hampshire Red. Intricate breeding programmes with these breeds has produced today's broiler chickens.

Housing

Specially constructed broiler houses are advocated. The use of non-specialist housing has been used to produce broilers, but the results are inferior to those produced under ideal broiler conditions. Controlled-environment houses are used in which the light and ventilation are regulated at the touch of a switch. The birds are brooded on the floor in unit sizes between 5,000 and 20,000. With such large numbers the ventilation must always be planned to meet the requirements of the maximum number of birds the house will hold at their greatest liveweight.

Thus, if a building 61 m x 6 m is used and each bird is allotted 0.5 m^2 and reared to 2.0 kg liveweight the ventilation arrangements must meet the needs of 5,000 birds whose maximum total weight will be 10,000 kg. The ventilation cannot be governed manually but by fitting thermostats into the wiring system. The thermostats cut in or out according to the minimum and maximum temperatures required in the building. A minimum ventilation is always provided.

The broiler house must be well insulated to maintain an even temperature throughout the year. Modern broiler houses have a U-value of less than 0.1 (this is a measure of its insulation value; the lower the figure the better the insulation).

Lighting the broiler

Many different lighting patterns have been advocated for broiler production, varying from continuous, 24-hour lighting to 12 hours' constant light in every 24 hours. Between these systems the variations are many.

It is usual to start the broilers with 23 hour natural lighting. The use of red lights reduces the incidence of cannibalism and also keeps the birds quieter, but the changeover from white to red lighting must be made gradually to avoid disturbance. Dim lighting, provided by 15-watt red bulbs, is constantly available even during the rest period; in this way panic is avoided. For optimum feed intake and growth rate at least 12 hours' daylight should be provided. This can be carried out in four 6-hour periods, two of which are bright light and the other two dim light. The procedure is one bright period and one dark period in every 12 hours. The dim light is provided by two 15-watt red bulbs for each 2,000 broilers and the brighter light by two 40-watt red bulbs for each 2,000 broilers. Blue lights should only be used for catching the birds at market time.

Brooding and management

Brooding is carried out by using mass-production techniques on the deep-litter system of management.

The litter should cover the floor to a depth of 8 cm, 1 tonne of wood shavings covering 186 m^2. Litter in the brooder must be covered with paper to prevent the chicks from eating the litter material. The paper may be removed when the brooder surround is 'opened up' after 4 days. House temperature should be 21°C and brooder temperature 34.6^{o}C. This should be reduced each week to a minimum of 21°C. Feed should be placed on egg trays or similar containers around the brooder's edge. Eight trays are needed for every 1,000 chicks. Water level should be provided in small glass jam jars or low-level automatic drinkers. Ten jam jar types are sufficient for 1,000 chicks. Fans should be set on minimum speed needed at the start of brooding.

When the chicks are 2 weeks old, five double-sided 2m drinkers will be required per 1,000 broilers. Twenty-five tube feeders are needed for every 1,000 birds. The feeders should be constantly adjusted in height to prevent spillage. Flint grit may be given in token amounts on top of the feed once per week. The litter may be raked over weekly, and any wet patches must be removed, replacing with new litter material. The brooders may be removed between the fourth and fifth week or suspended near the ceiling out of the attendant's way. Growers size flint grit should then be introduced, changing to an adult size at 8 weeks of age.

Feeding the broilers

Proprietary brands of broiler foods should be used by the smaller producer. The wide range of micro-ingredients required and the number of drug additives used in modern broiler production has made home mixing virtually impossible, except for the larger integrated business.

The food conversion rate and food cost per kg of liveweight is extremely important in broiler production. The food conversion rate is the amount of feed eaten in producing 1 kg liveweight. Broilers marketed at 8 weeks usually convert 2.2:1 whilst those marketed at 7 weeks will convert at around 2.0:1. Each broiler consumes during its

Two methods of raising table poultry. The birds are reared in cages until killing time. Some producers combine this system with floor rearing on the lines of deep litter, starting the growers in cages for a few weeks and then transferring them to a floor rearing unit

Seven week old broilers under modern conditions of housing and management

short, but productive, life between 4 and 4.5 kg of food.

During the first 28 days a starter crumb containing a coccidiostat and growth promotor should be fed. The food consumption over this period will be about 1 kg per bird. From 28 to 45 days a broiler grower pellet containing the same additives as the starter ration should be fed followed by a finisher ration to killing age.

High-energy rations are used. The energy is usually provided by cereals such as maize and sorghum but may be supplemented with oil or fat. Such cereals as maize are high in xanthophylls, which means they produce a yellow flesh. As the British housewife requires a white-fleshed bird the percentage of this cereal is strictly governed at a low level. Proteins of high quality are used in broiler feeding and the relationship between the metabolizable energy and protein is critical. A calorie : protein ratio of around 65 : 1 is used for the starter feed and around 75 : 1 in the finisher ration. Antibiotics may be used to stimulate growth rate and reduce the number of cull birds.

Coccidiostats are included in all broiler foods at a curative level. Their use is complementary to that of good husbandry for, if broiler producers attempt to rear their birds without good husbandry, outbreaks of coccidiosis are extremely likely. Common coccidiostats used are Pancoxin, Elancoban, Coyden, D.N.O.T.

Cleaning out the broiler house

Cleaning and disinfection of the broiler house should be done after every batch has been marketed. All remaining food should first be removed, all ventilators closed and the house fumigated with a 40% formalin solution. Microsol sprays may also be used.

After fumigating, the house is left for 12 hours before the equipment is removed for a physical clean up. The litter should be destroyed by burying or burning if no other outlet is available. The house should be brushed down and the floor scrubbed with a washing soda solution. The house should then be refumigated and the equipment reassembled in readiness for the new batch of chicks.

A period of at least 7 − 10 days must be allowed between cleaning out and restocking.

Broiler breeding

As with the production of table broiler chicken, broiler breeding is a specialised section of this industry calling for a high standard of

Broiler Meat Target Weights

Weeks	Weight lb	Weight kg
1	0.26	0.12
2	0.57	0.29
3	1.06	0.48
4	1.64	0.75
5	2.27	1.03
6	2.92	1.33
7	3.61	1.64
8	4.29	1.96
9	5.04	2.27
10	5.61	2.59
11	6.40	2.90
12	7.20	3.27

management and stockmanship. Broiler breeding is almost always carried out on a contract basis, the egg producer sending all his eggs to the hatchery which supplied him with the breeding stock. The breeding egg producer receives a standard price per 12 eggs, based upon current costs of production depending upon the hatchery concerned. These are generally basic prices, the producer receiving more or less depending upon the hatchability of his eggs over and above a certain percentage, e.g. a bonus may be payable for each 1% increase in hatchability over 80%.

Broiler breeding stock produce about 150 eggs per bird in a breeding period of 9 months; of these about 80% hatch into 1st quality chicks. Management must be good for the loss of only a few eggs through disease or negligence may mean a financial loss to the producer.

The breeding birds are large bodied and weigh approximately 2.3 kg at point of lay. Because of their size appetite is also high. Special broiler breeding foods are therefore provided by feeding-stuff compounders which are 'low' in energy. Overfatness must be avoided.

Most hatcheries supplying broiler breeding stock insist upon minimum flock sizes of 2,000. In this way administration and overhead costs are reduced, and the broiler chicks are produced more economically to the advantage of hatchery and broiler grower.

To obtain the high rates of production and hatchability necessary in the production of hatching eggs for the broiler industry management must be of a high standard. The pullets should be reared, in controlled-environment houses under the 'step-down' or constant daylight system of lighting. Each growing pullet must be allowed at least 0.18 m^2 of floor space and should be reared separately from the cockerels. Feeding should be in the form of a growers rations and fed in restricted amounts with oats or barley. Adequate feeding space must be available to prevent bullying and cannibalism. At least 10 cms of trough space should be allowed for each pullet from 14 weeks of age to the commencement of lay. The birds are put into their laying houses when 20 weeks of age. Culling is carried out at this age and any underweight pullets are re-moved. Debeaking may also be carried out. The breeding pens are mated up in the ratio of one cockerel to every ten pullets. It is usual to de-toe nail the cockerels to prevent them tearing the pullet's backs during mat-ing. The breeders should be fed a broiler-breeder ration with a scratch feed of grain in the litter. Oyster shell grit may be fed but only on the advice of the feed supplier.

Each bird should be allowed 0.2 - 0.25 m^2 of floor space, 15 cms of trough space and 3 cms of water trough space.

The hatching eggs are first collected 1 month after introducing the breeders ration. The eggs should be collected at least three times per day and stored at a temperature of 10 - 12.5^{0}C and R.H. of 80%.

Artificial caponizing by means of a female hormone pellet is no longer permitted under European Regulations.

Free Range Cockerels

A recent move has been the development of birds for free range, which gives a much better tasting chicken. This is to combat the allegations that intensive reared birds are tasteless. Statements that the meat tastes like card board and are lacking food value are now common place. How-ever, it must be appreciated that familiarity breeds contempt. Moreover, with the downward shift of prices it was inevitable that broilers had to be produced as cheaply as possible. The public want cheap chickens, but they do not always get the quality they expect.

There is also the problem of the growing awareness of the use of antibiotics and chemical additives in the food stuff. As a result, stricter controls are now imposed and meat offal is not permitted and other in-gredients have to be produced in a manner which reduces the danger of salmonella or other diseases.

Intensive Rearing of Birds. Note all aspects are automated –
feeding, drinking and air conditioning.

Crosses for Free Range Production

A cockerel that develops fairly fast and is quite hardy is the type which
should be selected for fattening. In effect, there is a return to the Sussex
type of fowl, which was a major industry before intensive methods were
used.

Because of the access to grass and other natural food the free
range pellets can be used, which means economy in labour costs. Alternatively, the farmer can do his own mixing using ground oats, barley,
milk and fat which was the traditional recipe. The oats and barley (8
sacks of oats and one of barley) were ground very fine for rapid digestion and then mixed with skimmed milk, and a small amount of mutton
fat (5 oz. per bird), until the mixture was crumbly (See *The Sussex &
Dorking Fowl,* J Batty, BPH).

The aim, which was achieved, was to produce very tasty chickens, which had lived a natural life up to the point of being killed. The
birds had white legs and flesh, with a broad breast.

The aim was to produce table birds at 12 to 16 weeks of age. Under the old system, cramming machines were used which meant that birds were force fed by means of a funnel or pipe for the last two weeks or so, thus putting on extra weight. This practice is no longer permitted, but the same effect can be obtained by restricting the birds movements in a pen so weight is added by giving more food.

Breeds which have been successful on free range are:

1. Sussex which is a dual purpose breed yet quite an acceptable layer as well.

2. Indian Game X Light Sussex. This gives extra weight, although some may mature more slowly than 1.

3. Sussex X Wyandotte. This gives more vigour and should produce white flesh.

Recently the so-called "Corn Fed" table bird has been marketed which is fed on pellets and maize which increases the yellow pigment on the skin. Whether this will be successful is not certain because the British public prefer the white skinned birds.

Other breeds such as the Orpington and Dorking may also be used, but these should be a strain which develops fairly quickly. Some breeds produce very tasty table birds, but are not economic to produce.

If table birds are to be produced there may be a case for using the Silver and Gold colours of the Sussex, which allows males and females to be segregated at birth. The Light Sussex Male (Silver) crossed with Red Sussex Females ((Gold) will produce red males and silver females. Details on how the system works is given earlier (pages 18 and 19).

Chapter 8

TURKEYS

INTRODUCTION

The consumption of turkey meat has shown considerable increase since the early post-war years. Until a few years ago, turkey meat was regarded as a luxury and the purchasing public were prepared to pay a good deal for a meat invariably purchased only once a year.

Since these times expansion within the turkey industry has been rapid until, at the present day, something like 20,000,000 birds are produced and consumed annually. Half of this production is taken-up at Christmas, and already new markets are being created at other times of the year, notably Easter and Spring Bank Holiday. Indeed, the main aim of the now-established British Turkey Federation (B.T.F.) is to promote and sell turkeys all the year round. Turkey production has now become a highly specialised branch of the poultry industry accounting for some 10% of the nation's poultry population.

Market requirement falls into two classes. First, there is the traditional Christmas trade. The size of turkey demanded at this time of the year is on the average considerably larger than that marketed in the interim periods. Weights are from 6 kg oven-ready upwards. No upper limit is given because many birds are used in the catering trade. Secondly, there is the all-the-year-round turkey trade. Oven-ready weight ranges appear to be far lower and narrower; most popular weights are 3 − 5 kg. This chapter deals with the techniques used in the production of these two classes of turkey.

Broiler or teenage turkey production

Large scale turkey production on free range. Note the bulk feeding hopper in the background

Breeding

Turkey hens are not prolific egg producers and on average lay between 90 and 120 eggs each in the breeding season. The lighter type of turkey produces at the higher rate, the heavier type at the lower rate. A target of saleable poults/hen is between 50 – 70

Characteristics such as egg production and hatchability have a low hereditability, which means that improvement of them is dependent upon family selection and progeny testing. Records of these are obtained by trapnesting the hens. For satisfactory progress a minimum of 100 breedings hens is required. These should, in the case of natural matings, be split up into groups of ten and mated with one stag. Careful recording will show the superior families and allow accurate comparisons to be made between individuals. Recording systems must not simply take into account the breeding potential but also the rearability of the offspring. With larger breeding flocks strain crosses are used in the production of hybrid turkeys suitable for the specific market catered for.

The normal breeding season is between 4 and 5 months. Larger breeds and strains are usually active only for the shorter period, while the lighter strains remain active over the longer period.

Selection programmes will usually be based upon the following economic characteristics: egg production, hatchability, liveability, shell texture and body conformation. Good shell texture is important, for the breeder must attempt to obtain as many settable eggs as possible. A.D.A.S. geneticists give advice on genetic programmes.

Mating

With the trend towards shorter shank length and broader breast width in market turkeys the breeder is faced with mating problems in breeding birds selected for extreme broadness or dimple breastness. Stag turkeys exhibiting extreme breast widths are not able to perform satisfactory natural matings, the bird losing its balance before completing the act. In such circumstances artificial insemination, either completely or partly helps in maintaining fertility. Breeders of very broad-breasted turkeys use only A.I. and do not rely upon natural matings at all. Turkeys with less-extreme breast conformation may be used for part natural and part A.I.

As turkey semen cannot be stored for more than about 30 minutes the stags have to be 'milked' immediately before inseminating the hens.

This is done by hand manipulation of the area surrounding the stag's cloaca. The hens are inseminated once every 7 — 14 days using a special syringe. The volume of diluted semen used is 0.05 cc per hen. On modern turkey breeding farms both hens and stags are often housed in turkey cages. This system of management and housing reduces the time wasted in catching the hens. They can be inseminated without removing them from their cages. A.I. is also used when valuable males are removed from a breeding pen owing to injury, etc.

In natural mating the ratio of hens to stags is usually 8 — 10:1, depending upon the virility of the stag.

Housing turkey breeders

Turkey breeders are housed in a variety of ways, one of the most popular being in fold units measuring 6 m x 1.52 m. Under this system the hens are folded over grass leys. Where family or sister groups are housed in one fold unit trapnests are used. Straw or poleyards are also a popular method of housing. In these circumstances flock mating is usually carried out, trapnesting being of little use as it is impossible to know which stags sired the eggs.

Polesheds, which are also used extensively for fattening, consist of open-sided units constructed of wire netting, poles, corrugated asbestos or three-ply felt. Ventilation troubles are non-existent, and artificial lighting can be provided when required.

Lighting turkey replacement breeding stock

Turkey breeders can be brought into production irrespective of the season of year by using artificial lighting. Hens respond to increases or decreases in daylength far more abruptly than the stags. For this reason, stags should be lighted some 2 — 3 weeks before the hens are required to lay. Breeding replacements may be prevented from laying too early be rearing them under normal daylength conditions until they are 16 — 18 weeks of age. Decreasing the total daylength to about 8 hours in each 24 hours prevents early sexual maturity and consequently the production of small hatching eggs. This holding period should be maintained until 3 — 4 weeks before eggs are required. Increasing the light at this stage will stimulate egg production. Stags must not be subjected to this lighting pattern but reared under natural lighting conditions until 5 — 6 weeks before hatching eggs are collected.

Incubation of turkey eggs

In the main the principles of incubation outlined in Chapter 2 apply equally well to turkey eggs. However, as turkey eggs have an incubation period of 28 days compared with 21 days for chickens there are a number of differences which must be mentioned.

Eggs of small type will weigh 70 − 85 g and those from large types 85 − 100 g. Only fresh eggs of good shape and shell texture should be incubated. As a means of controlling bacterial contamination the hatching eggs should be dipped in a suitable germicidal solution for 15 minutes at a temperature of 29.4°C. If possible, fumigation with potassium permanganate and formalin should also form part of the pre-incubation practice either on the farm or in the hatchery.

Special turkey incubators should be used. Those used for hatching chicken eggs are not always suitable owing to the differences in egg size. Manufacturer's instructions must be followed. The following incubation temperatures should be used as a guide for hatching turkey eggs in cabinet-type machines.

Period in Days	*Temperature in °F*	*in °C*	*Relative Humidity as %*
1 − 24	99.5	37.5	60
24 − 28	97	36.1	75

The eggs should be tested for fertility on the 24th day and transferred to the hatching trays. Hatching will commence on the 26th − 27th day and should be completed by the end of the 29th day. Depending on fertility, the hatchability of all eggs set should be about 75%. Hatchability of fertile eggs should be around 90%.

Brooding and rearing the poults

The standard period of brooding is regarded as 6 weeks. In the summer, however, 4 − 5 weeks will be sufficient. Brooding can be carried out in many of the brooders discussed in earlier chapters, but it is important to remember that, comparatively, turkey poults are larger than chicks and require proportionately more space and headroom. Brooders capable of holding 100 chicks will brood only 50 − 60 poults to the same age. Tier brooders used for chicken rearing are suitable for turkeys up to the age of 2½ weeks. After this they must be transferred

Folds are popular for breeding turkeys, each taking eight to ten hens and a stag. They are moved daily and littered with straw in bad weather. Turkeys should not be run with other poultry, nor put on land over which such poultry have ranged until at least a year afterwards

These turkeys are being raised in a compound, formed by an old farm building with straw yard attached

into a bigger brooding area. Special turkey brooders are available and where possible should be used.

On specialist turkey farms intensive floor brooding is extremely popular, batches of 5,000 or more being brooded under one roof on similar production lines as those used in the chicken-broiler industry. Labour is reduced to a minimum and brooding costs are lower. The following amounts of floor space are recommended for intensive floor rearing.

Age in weeks	Floor space in sq. metres
0 – 4	0.07
4 – 8	0.14
8 – 12	0.23
12 – 16	0.27
16 – 20	0.37
20 – 24	0.47
24 – 28	0.56
28 – upwards	0.65
Breeders	0.65

In the production of so-called 'teenage' turkeys, i.e. birds marketed between 12 and 18 weeks of age, specialist controlled environment housing is necessary for maximum body weight gain and food efficiency. Well-insulated broiler-type housing has many advantages, especially on turkey farms practising all-the-year-round production. Ventilation must be good and may be calculated according to the standards given in Chapter 5. Feed- and water-trough space must be adequate, so that all the turkeys can feed at the same time. The following feed comparison and bodyweight table may be used as a guide. Figures are cumulative per 1,000 poults., expressed in kilos and pounds.

Turkeys

Age in weeks	Average body weight	
	lb	kg
1	250	113
2	750	340
3	1,500	682
4	2,200	1,000
5	3,600	1,636
6	5,500	2,500
7	7,500	3,409
8	10,000	4,545
9	13,000	5,909
10	16,000	7,272
11	19,500	8,863
12	23,000	10,454
13	27,000	12,272
14	31,000	14,091
15	36,000	16,363
16	39,000	17,727
17	44,000	20,000
18	49,000	22,272
19	55,000	25,000
20	60,000	27,272
21	65,000	29,545
22	73,000	33,181
23	77,000	35,000
24	83,000	37,727
25	90,000	40,909
26	100,000	45,454
27	104,000	47,272
28	112,000	50,909

Because the stocking rate is high in 'teenage' turkey production safe-guards must be taken against outbreaks of cannibalism. Low lighting intensities are used with success especially when red bulbs are installed. An intensity of 0.05 lux is sufficient for the poults to see to eat and drink in comfort, and at the same time it deters cannibalistic tendencies. White light may also be used, but blue light must be avoided for rearing. The intensity of light recommended applies to turkeys 6 weeks of age and over. Poults up to this age and especially during the first 2 weeks of life must receive high-intensity lighting to encourage them to start eating. Feeders and water drinkers must not be situated in dark corners of the house. This will lead to so-called 'starve-outs' caused by insufficient food consumption. Perches are not advised in the production of this type of turkey as they may predispose to breast blisters and crooked breast bones. Litter must be maintained in a dry condition.

The production of the more traditional type of turkey requires less-specialised housing. Requirements of brooding up to 6 – 8 weeks of age are, however, identical. After this age various management systems can be adopted, varying from free range rearing to strawyard or pole-shed rearing. Turkeys over 8 weeks of age are extremely hardy and can put up with comparatively wide fluctuations in temperature. Only simple housing is required.

Production in strawyards

This system is most popular on the general farm and is suitable for large- or small-scale production. It often makes use of existing housing and surplus straw and thus often fits into a general farming programme. The use of large self-feed hoppers and automatic waterers is labour saving and economic. For turkeys marketed at Christmas when they are 22 – 24 weeks old 0.46 m2 of yard space should be allowed each bird. Groups of several 1,000 may be safely reared together, although the inexperienced poultry man will find 500 bird groups more manageable. The strawyard should be well drained to prevent worm infestations. A gravel or sand base is ideal.

Production in verandahs

A popular method of rearing small groups of turkeys in verandahs is called the Motley system. Each verandah unit, which measures 6 m long, 1.5 m wide and 0.75 m to the eaves, has a capacity of 30 turkeys to 20 weeks of age. The verandahs are constructed of timber and wire netting. The floor consists of wooden slats 2.5 cm apart, made of 5 cm x 1.25 cm planed timber. Each unit has an apex roof covered in timber and felt. As the floor of the verandah stands 0.75 m from the ground an accumulation of droppings can occur, thus allowing a once-a-batch clean out. Feeding and watering is provided by troughs sited around the sides and on the ends of the unit. The system is more suitable for producing traditional-type birds in the summer months, but because of heavy labour demand is unsuitable for large scale production.

Free range

Where space permits, turkeys can be satisfactorily reared under free range conditions. Labour requirements tend to be higher because usually both feed and water have to be carted to the birds. Strong, high fences are necessary in fox-prone areas, and careful watch is necessary at marketing time in the event of poachers. On good pasture the food requirements may be reduced by as much as 10% although the growth rate will probably be proportionately slower.

Separating the sexes

Separating the sexes either at day-old or later in life results in a more economic production of both males and females. If not done at day old it can be carried out at about 10 weeks of age. There is less fighting and treading in sexed poults and the percentage of grade A birds is generally improved. Because hens have lower nutritional requirements than stags, feeding can be tailored to meet the demands of both sexes. Male turkeys have a 'redder' and more pronounced head parts than females. They also show courtship display from an early age.

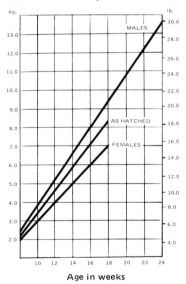

Turkey – Liveweights

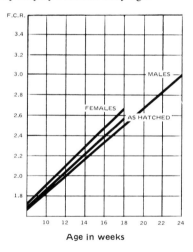

Turkey – Food conversions

*Caponisation**

The 'finish' or 'bloom' on 'teenage-type' turkeys can be considerably improved by using synthetic oestrogens either in the turkey's diet or by implantation. This fact is extremely important to producers and retailers. Caponising can be carried out by using either stilboestrol or hexoestrol implants. The usual dosage level is 15 mg per poult given at about 4 weeks before killing. The implant is injected below the skin at the base of the neck region, not into the flesh, by using a small syringe. The effect of chemical caponisation is to increase the deposition of subcutaneous and abdominal fat, thus making the carcass appear creamier in colour and juicier to eat. Caponisation is not necessary in more-mature birds and must on no account be used for breeding stock.

Marketing

Turkeys bruise extremely easily, especially those produced at an early age. Therefore, they must always be handled with extreme care. When catching and crating only the shanks should be held, not the turkey's drumsticks or wings. Damage to the skin detracts from the appearance and will result in downgrading by the packing station. The period of starving prior to killing should be 8 − 10 hours. Water must, however, be available throughout this period.

Killing in packing stations is usually carried out by stunning and bleeding. In this way a white flesh is obtained. Killing by dislocation is only practised by small producer-retailer turkeymen. This is done by hand or with the aid of a broom stick. The bird's head and neck are dislocated leaving a cavity for the blood to drain from the body.

Plucking can be done by hand, in the case of the small producer, or by using one of the small dry-plucking machines. Large packing stations specialising in processing turkeys pluck by the wet method (semi-scald method). The birds are immersed in water at a temperature of 54.4°C for about 30 seconds. This process loosens the feathers and allows the automatic plucking machines to carry out their work efficiently. Modern plucking machines strip the feathers by means of rubber fingers. Stubs or pin-feathers are removed by a further set of fingers. Wax plucking is occasionally used but the cost is generally prohibitive.

During killing, plucking and processing a certain amount of body weight is lost. The actual amount will vary from strain to strain, and the following figures must be used only as a guide:

* No longer practised in EEC countries.

Turkeys

Loss through starvation	2%
Loss through killing and plucking	8%
Loss through evisceration	15%
	25%

The younger the bird the greater the loss and vice versa.

Chapter 9

TABLE DUCKLING PRODUCTION

INTRODUCTION

The production of table duckling is as highly specialised as other sections of the poultry industry. The market is limited and demands that only first-quality white skinned duckling weighing between 2.5 and 3.5 kg should be marketed; they must also be light in bone and carry an abundance of breast meat. The pure Aylesbury duck fulfils most of these conditions but as a breeding bird has the disadvantage of a low annual egg production. To a certain extent modern breeding methods have altered this pattern by crossing this breed with the Pekin or White Campbell. As these latter breeds produce yellow flesh the straight cross-bred does not always command the maximum return and hybrids have therefore been developed.

Duckling with dark plumage are unsuitable as they produce dark stub feathers which may stain the carcass. Table duckling are marketed when 7 – 8 weeks old and before they start growing new feathers. When this occurs it is often necessary to allow the ducks to complete the feather growth and market them at 14 – 16 weeks of age. This is not economical and is rarely practised.

Breeding stock

Usually only first-year ducks are used in table duckling production. This ensures a greater and earlier supply of fertile eggs, thus reducing overheads and therefore the selling price of the day-olds. For the production of breeding stock replacements second-year ducks are used with first-year drakes. The mating ratio for Aylesbury is one drake to

Infra-red brooding can be used for ducklings as well as chicks. The wire surround keeps the birds near the source of heat for the first few days. Wire floors, however, are better than solid ones for ducklings, as they help to keep them drier and happier

every four ducks. For Pekin and Hybrids one drake to each five or six ducks. When flock mating is practised 25 or 30 ducks are mated to six or seven drakes. As the season progresses the number of drakes may be reduced by 15 – 20%, for, as the ducks cease laying so the drakes become less active. The average number of day-old duckling produced by each duck is around 50 – 55 for Aylesburys and 60 – 65 for Hybrids. The breeding stock should be selected on growth rate, food conversion, liveability, egg production and hatchability. First, selection is usually carried out at 6 – 7 weeks of age on body conformation and feed efficiency. Where the ducklings are wing banded to the sires record it is possible to select on food conversion rate, hatachability, etc. The second selection is carried out just before housing, at point of lay, when unhealthy ducks are removed.

As the natural laying season of breeding ducks occurs during the spring and early summer – and the best prices for market duckling are at this time also – it is necessary to light the breeders artificially so as to obtain hatching eggs at the most profitable time. Since it takes some

3 months from the time the egg is laid until the duckling is marketed at 8 weeks it is necessary to plan the breeding season well in advance. By using artificial lighting the breeding stock can be brought into lay during the early winter months. The ducks should be 20 – 24 weeks old before they are lit. Laying commences about 14 days after the extra light has been introduced. The drakes must be lighted 2 – 3 weeks before the ducks to ensure that the first eggs layed are fertile. Peak production is generally reached 8 weeks after laying has started. A second peak is occasionally obtained 7 months after the first but the aim is to produce consistent high production!

In large breeding concerns it is necessary to produce duckling economically all the year round. This is done by using three or four breeding flocks and bringing them into lay so that throughout the whole year one flock is laying at its maximum. With smaller breeders the same results may be achieved by lighting only three-quarters or two-thirds of the flock at any one time. Continuous production may not be obtained, but the peaks and troughs in production are not so severe when compared with results obtained by lighting the whole flock at one time only.

BROODING AND REARING THE DUCKLING

In comparison with other farm animals the growth rate of Aylesbury duckling is phenomenal, as the following figures show:

Age in weeks	Weight lb	Weight kg
2	14　oz	0.39
4	2½ lb	1.13
6	5¼ lb	2.38
8	6¼ lb	2.83

To help achieve these weights the initial brooding period must be ideal. Tier brooders can be used up to 3 weeks of age. After this age the duckling quickly will outgrow this type of unit. No artificial heating is normally necessary after 3 weeks during the spring and summer months. In winter time the brooding period may need to be extended to 4 weeks. Duckling are messy drinkers and the tier brooders may quickly become unpleasantly dirty. Ventilation of the brooder house should be extremely good to deal with the high humidity and drinking points should be sited over a wire floor area.

Floor space requirements in solid-floor units

When too closely confined the duckling will peck each other, often stripping the 'down' from each other's backs. Tier brooders 2.75 m x 1 m will hold 60 duckling to 3 weeks of age.

Age in weeks	Area in m^2 per bird
0 – 2	0.046
2 – 4	0.093
4 – 6	0.186
6 – 8	0.279
8 – 10	0.371

Extensive systems of rearing are practised in fox-free areas and where plenty of sandy or free draining land is available. The duckling are brooded intensively for the first 2½ weeks and then transferred to open range. Wire netting 0.5 m high, is adequate to confine them. Sleeping quarters may consist of curved corrugated iron sheets or straw baled compounds. The biggest disadvantages of this system are the high feed wastage by wild birds and losses from vermin. Swimming water is unnecessary for fattening ducklings (Fig. 9.1).

Feeding table duckling

Duck starter pellets or crumbs containing 18 – 19% protein should be fed for the first 3 weeks. This should be provided *ab lib*. A duck fattening or finishing ration should be fed from 3 weeks to killing. A ration containing 16% protein is adequate in this period. Allowance must be made for the duckling to feed and drink at night as they are natural nocturnal feeders and grow faster when allowed access to feed 24 hours a day. Wet-mash feeding can be practised with small lots but is not really advised, for not only is it labour consuming but the risk of stale food may cause digestive upsets. Antibiotics are of little benefit to duckling probably because the content of the gut is so fluid. Coccidiostats are not usually provided.

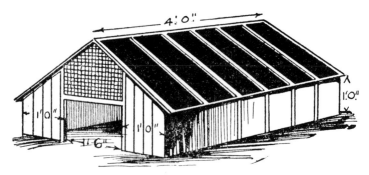

Fig. 9.1. Duck housing can be very simple in nature. Here is a small and easily made portable shelter, suitable for either adult or growing ducks.

The following figures are a guide to food consumption per duckling.

Age in weeks	Weekly in kg	Cumulative in kg
2	0.77	0.77
3	0.85	1.62
4	1.22	2.84
5	1.36	4.20
6	1.50	5.70
7	1.52	7.22
8	1.59	8.81
9	1.54	10.35
10	1.47	11.82

There is very little difference in growth rate between the sexes and therefore there is little benefit to be gained from separating the ducks from the drakes.

KILLING, PLUCKING AND EVISCERATION LOSSES

Killing is carried out at between 7 and 8 weeks of age when the flight feathers are approximately 8.9 cm long. The difference between young and old birds is determined by feeling the windpipe. It is soft and pliable in young birds and hard in older ducks.

Killing is best done by cutting the jugular vein. Bleeding makes the carcass white. Dislocation of the neck may be carried out but does not produce such a white carcass.

Plucking duckling is best done with hot wax. This removes all the down and stubs. It has the disadvantage, however, of wasting the valuable down feathers. To recoup some of this loss dry-machine or hand plucking may be used first followed by wax dipping to impart a 'bloom' on the carcass.

After rough plucking the carcass should be allowed to cool for 20 – 30 minutes before wax dipping. This allows the body temperature to decrease and increases the 'binding' properties of the wax. The temperature of the wax should be 60°C. After wax dipping, the birds

White fleshed and well-feathered, the Aylesbury is the outstanding duck breed for table purposes

are immersed in a cold water bath to harden the wax. This is then stripped off 5 – 10 minutes later and the wax reclaimed from the feathers.

The loss of weight caused by killing, plucking and evisceration varies considerably. Bleeding results in a weight loss of 3 – 6%, plucking 5 – 7% and evisceration 28 – 32%. These figures are percentage loss of liveweight.

Duckling may of course be wet plucked in a similar manner to chickens. The feathers and down can be reclaimed but are less valuable compared to duckling dry plucked.

Duck Egg Production

The table duck has long since been a favourite, but the possibility of duck egg production is largely neglected. This is partly due to a prejudice against duck eggs, yet they are quite nutritious and good for baking. Moreover, in terms of production the Khaki Campbell and Indian Runner ducks lay far more eggs than a fowl, and are very hardy.

Housing can be along the lines shown earlier in this chapter, and layers' pellets and mixed corn in water is the normal food. Make sure that the birds are locked up at night and not let out too early or the eggs will be laid outside.

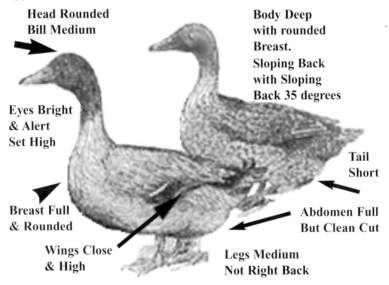

**Head Rounded
Bill Medium**

**Body Deep
with rounded
Breast.
Sloping Back
with Sloping
Back 35 degrees**

**Eyes Bright
& Alert
Set High**

**Tail
Short**

**Breast Full
& Rounded**

**Abdomen Full
But Clean Cut**

**Wings Close
& High**

**Legs Medium
Not Right Back**

Khaki Campbell Pair – showing requirements for exhibition.

Chapter 10

GEESE

INTRODUCTION

Profit from geese comes mainly from sales at Christmas time as they are one of the varieties of poultry meat acceptable for the festive season. On the other hand, profit can be made from the sale of goslings or hatching eggs. Goslings reared for selling for future fattening can be sold as early as 1 week old up to 10 – 12 weeks. There is always a demand for such goslings, especially to farmers who have grassland suitable only for grazing.

Table prices are rarely generous and therefore total production costs have to be kept as low as possible. Fortunately, geese are exceptionally good grazers and they will eat far more grass, weeds and herbage than any other type of poultry. For this reason, they must be encouraged to eat and range over grassland. For good weight gains the grass should be of good quality. They are ideal for grazing over orchards or where the owner wishes to keep the grass short without expenditure of labour. Geese are excellent 'watch – dogs'.

BREEDS

The main pure breeds used in the production of table geese are Embden, Toulouse, Roman and Chinese. Hybrid geese are also available and generally given superior breeding and growing performance to pure breeds.

The Embden is a most useful variety of goose, answering most useful purposes and giving exceptionally good results in most situations. It is

These geese have been put into an enclosure in the farmyard to finish them for market

Toulouse geese, a heavy breed that grows well and is useful for crossing, particularly with the Embden. Although extremely hardy they should be given some shelter and protection from rain during the winter to encourage early laying

snow white in colour and has plenty of good, plump breast meat. It usually lays about 30 eggs in its first season. It weighs 9 – 14 kg. depending upon sex, and is often used for crossing with the Toulouse.

The Toulouse is slightly smaller than the Embden but is a useful meat producer. Egg production is around 40 eggs per bird in the first season.

The Roman is a white goose, light in bone without a lot of offal. They are good layers of medium-sized eggs, some strains laying up to 75 eggs in their first season. It is a quick-maturing goose weighing between 5 and 6 kg.

The Chinese goose is a small bird, hardy and very fertile. It is a prolific layer often laying up to 90 eggs per year. The Chinese has a dark flesh and is often used for crossing with other varieties to improve egg production. Today, hybrid geese are produced both in the U.K. and on the continent.

Breeding and mating

Geese should be 6 – 8 months old before being used for breeding. Only sound, healthy, vigorous stock should be used. Geese of good weight should be selected, making due allowance for how they have been fed during rearing. When first-year geese are used they should, for preference, be mated with a second-year gander. This ensures that the young goslings will be strong and healthy. The gander should be active and alert. Geese are not always successful breeders during their first year but tend to improve with age up to 10 years or even older. A breeding pen is called a 'set' and usually consists of one gander to every three of four geese. With Chinese five geese may be used. Geese not invariably partner for life.

Young geese commence laying in February and cease about the middle of June; late-hatched birds may continue laying into summer. Geese must be mated up well in advance of anticipated egg production to obtain good fertility. Early autumn mating is recommended to allow the birds to become settled and contented in their surroundings.

It is not advisable to let the geese do their own hatching. By and large, they make poor setters. Further, to obtain maximum egg production non-broodiness is essential. Broody geese should be put into a small wire-netting run in full view of the other geese. They should not be allowed to make a nest. Frequent egg collection helps prevent broodiness.

It is essential to feed a breeders diet to breeding geese to obtain good fertility and hatchability. This food should be introduced to the 'set' at least 1 month before hatching eggs are collected. During the breeding season no other food should be fed. Later in the season grain up to 14 g per bird may be fed. A little oyster shell grit must also be made available.

The eggs should be collected each day. In frosty weather they must be collected directly they are laid. Dirty eggs may be brushed clean with steel wool. They should be turned daily until required for incubation. Stale eggs hatch less well than fresh eggs and rate of hatchability declines rapidly in eggs more than 10 days old.

Incubation

The incubation period varies from 28 to 30 days. After 26 days 'chipping' may be noticed, although it may not be completed until 48 hours later. Hens make good mothers, a large broody being capable of dealing with four or five eggs. The soil around the nest should be moistened daily to keep the humidity at the correct level. Natural-draught incubators are often used successfully for hatching goose eggs. With this type of machine it is necessary to raise the thermometer bulb to 6 cm above the eggs. In this position the temperature should be 40°C, in cabinet or forced draught incubators the correct temperature is 38°C. The eggs should be turned three or four times each day. Eggs under broody hens should be turned once a day as they are too large for natural turning.

Brooding and rearing

Most chick brooders are satisfactory for rearing goslings, provided it is remembered that manufacturers' recommendations for chick capacity should be reduced by 75%. Brooders manufactured for duckling are particularly suitable for goslings. The best feed is chick crumbs or pellets for the first 2 − 3 weeks. Food and water should be provided *ad lib.* Water troughs must be sufficiently deep for the young gosling to immerse its head fully. Any observant rearer should be able to ensure the correct brooder temperature and an ample supply of fresh air. As a guide, the brooder should be started at about 35°C and reduced daily until at 3 weeks of age the goslings are without heat.

In warm weather heat may be dispensed with by 2 weeks of age. At 3 weeks the goslings will be ready to be moved into a protected run and housed in a wire-netting or solid-floor ark or outdoor brooder. They will quickly make use of the good, short grass and should be moved regularly on to fresh ground. After 3 − 4 weeks a growers ration can be given in small quantities at the end of the day. Mixed corn may be given by 8 weeks in small quantities and may completely replace the growers ration by 10 weeks of age. Growing goslings must be given protection from the sun and rain: an orchard is admirable. Swimming water is unnecessary for fattening geese and, with few exceptions, is usually unnecessary even for breeding geese. Water deep enough for head immersion is, however necessary.

The sexes can be distinguished at quite an early age without much trouble. The bird is held securely but gently on its back. The tail is folded under the back and the anus is exposed. With the first two fingers of one hand use a gently but firm outwards and downwards pressure. This will expose the genital organs which in the case of the male will show as a fleshy 'pencil' while that of the female will be concave and smooth. The novice should enlist the help of an expert before attempting the operation. Other indications of sex differences are external factors. In the male or gander the neck is usually longer and thicker than the female's. The legs, too, are generally thicker and stouter in the gander. By using the methods described above an accuracy of 90 − 95% should be quickly attained.

Goslings reared in fox-free areas will require only limited shelter. Simple lean-to, straw-baled constructions are generally adequate, for it will be found that the geese will spend almost all their time outside. Geese are naturally nocturnal feeders and for maximum weight gains they should not be confined at night.

In areas where foxes are a nuisance it will be necessary to confine the geese. The compound should be enclosed with 2 m high wire netting buried at least 0.3 m in the ground. The house must, of course, be of a more permanent construction to prevent the geese from escaping and foxes from entering. In these circumstances both food and water should be provided in the house (Fig. 10.1).

Geese are not difficult to control in ordinary circumstances and, even if they do wander, they usually return towards evening time once they are accustomed to their surroundings. However, many of the lighter breeds, such as the Chinese and Roman can fly well and for this reason they should either be pinioned or have one wing clipped. The

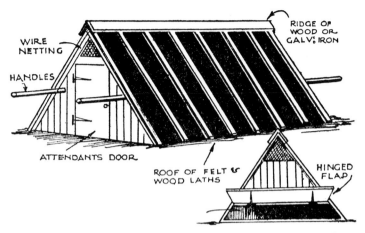

*Fig. 10.1. A useful apex type goose house suitable for breeding stock.
Its portability is an added feature of its utility*

cutting of the flights only lasts until the birds acquire new feathers, when they must again be clipped.

Fattening geese

The gosling can usually be left on grass until November, when, for maximum weight gain, they should be brought into a confined area, such as a strawyard. Under these conditions each bird should be allowed 0.75 m^2 and fed on concentrates. A simple fattening ration containing about 14 – 15% crude protein should be fed with equal parts of grain, such as barley. At each feed it should be given as much as it can clear up within an hour of being fed. Swimming water is unnecessary, but sufficient should be given for complete head immersion.

AILMENTS AND DISEASES

Geese are very strong, hardy birds, generally spending most of their time in natural surroundings and living on non-forcing feeding stuffs.

Worms

Worm infestations are not uncommon in geese. The affected birds do not put on flesh to the same extent as those not infected. The geese should be moved on to fresh grass and treated with an antiworm remedy, such as carbon tetrachloride.

Embden geese make an excellent cross with Toulouse for the production of Christmas birds. Birds that are healthy, grow rapidly and fatten cheaply, are essential for successful goose production

Geese are ideal for grazing over orchards or where it is desired to keep the grass neat and short

'Slipped wing'

In cases of slipped wing the flight feathers protrude at right-angles to the bird's body; it may affect one or both wings. Affected birds should never be used for breeding purposes as the condition is almost certainly inherited. In young goslings it is possible to clip the flight feathers back or amputate the tip of the wing which supports these feathers. There is no known cure.

Lameness, swollen foot or bumble foot

This condition is most regularly seen in breeding stock where swimming water is absent. It is often caused by a staphylococci bacteria and can be cured by injection of antibiotics. Sometimes the core can be removed by cutting the swelling with a sterilised blade and removing the hard cheesy material. The affected birds should be isolated for a time to allow the wound to heal.

Coccidiosis

The goose is affected by the 'renal' form of coccidiosis between 3 and 12 weeks of age. The disease affects the kidneys causing high mortality. Affected birds become emaciated and may be found to lag behind the rest of the flock. The birds should be treated with sulphamezathine and moved on to fresh ground.

Goose influenza

This disease causes high mortality. Symptoms are loss of appetite and distressed breathing. Affected birds die within 2 – 5 days. There is no treatment for this disease.

Chapter 11

QUAIL AND GUINEA FOWL

INTRODUCTION

Although quail production is very much the cinderella of the 'poultry' meat market it is of sufficient importance, and will likely play an increasingly important role, to be discussed in some detail.

Originating in Japan or China, in the 11th century, quail have only in more recent years been used for meat and egg production. In these far eastern countries, the quail industry is large and in the U.S.A. it also is of commercial significance. The U.K. quail industry is small, estimated at under one million birds per annum. This level of production leaves room for expansion, but extreme care is needed in securing the market before either markedly expanding in the market or considering entering it.

Breeding

The Japanese quail is the most commonly kept stock. Bob White quail are not used commercially.

Two or three females are mated with each male and this can be in trios or flock mated in groups of 40–50 females. Wider mating ratios reduce fertility whilst over-mating has a similar consequence. Female quail breed for some 10–12 months and lay up to 300 eggs each in this period. The replacement of the males at the half way stage has a marked effect in boosting fertility. Egg laying commences when females are 10–12 weeks old. Artificial insemination is a possibility but not as yet used commercially.

Fertility from naturally bred stock is in the order of 86–90% whilst the hatchability of fertile eggs is 70% of eggs incubated. Inbreeding should be avoided where possible as it causes lowered reproductive performance.

Incubation

The incubation period is 18 days, but as with other birds, can vary by about one day either side. The ideal temperature is 100°F (37.7°C), but a suitable incubation regime is as follows:

1–12	days	37.7°C	at	58%	RH
13–16	days	37.0°C	at	53%	RH
17–17½	days	37.5°C	at	90%	RH
17½–18	days	37.5°C	at	53%	RH
(for drying off)					

Hatching itself is generally very quick and prolonged hatches are the exception rather than the rule.

Both cabinet and natural draught incubators can be used for incubation. A capacity of 5000 hen eggs is equivalent to 7000–7200 quail eggs.

Quail eggs should be collected several times each day and set before being 5 days old for good hatchability. Turning eggs in incubation is important and should not be less than five times each day.

Rearing Young Quail

Either cages or floor brooding may be used, both systems being good with an almost equal number of advantages and disadvantages to each system.

Cage floors should be covered with hessian in the first 2–3 days but this is usually unnecessary thereafter. The initial brooding temperature is 35°C reducing to 21°C by 3 – 4 weeks of age and being retained at this until killing at 6–7 weeks. Mortality to 7 weeks should not be greater than 3–4%. As the young quail chicks are so small, all gaps in the equipment must be stopped to prevent escape and water founts used for normal chicks should have pebbles or similar items placed in the trough to prevent drowning.

Approximately 30–35 quail chicks can be reared in each square foot of floor (30 square centimetres per chick). Cannabalism occurs if the birds are overcrowded, but can be controlled by using bulbs of low light intensity; e.g. 15 watt.

Quail housed on litter for table production (Photo: *Poultry World*, reproduced with kind permission).

Management of Breeding Quail

For best egg laying results the house temperature should not be less than 18°C and day length 17 hours in each 24 hours. The optimum relative humidity is 60% as this helps to reduce dust levels and control respiratory problems.

Cage breeding is popular and each breeding quail needs 180 cm^2– 200 cm^2. Flightiness is controlled by maintaining cage height at 15–16 cm (6½"). For floor/litter maintained breeders, 360 cm^2 is required. Lighting is an important environment aspect. Only 8 hours in each 24 hours is necessary for growing and fattening stock. Light intensity should be below 1.5 lux in order to control flightiness and cannabalism. After 8 weeks of age the breeders can be changed to 14 hours and increased to 17 hours by around 16 weeks, thereafter, being maintained at this day length. The males should be held at 14 hours until 12 weeks when they are mated and naturally changed to the female programme.

Feeding Quail

In many respects quail have similar nutritional needs as turkeys. That is, in the early growing period high protein rations are necessary and in the breeding stage both high protein and high levels of vitamins and minerals are essential for fast growth and good reproduction respectively.

Recommended protein and energy levels are as follows.

Age Days	Protein %	Metabolisable Energy	
		kcals lb.	MJ/kg
1– 7	30.0	1300	12. 0
8–21	26.0	1300	12. 0
22–42	20.0	1320	12.17
Breeders			
84 days–300 days	20.0	1280	11.80

Growing quail need 1.0% and 0.8% calcium and phosphorus respectively whilst the quail breeder ration should contain 2.75% calcium and 0.75% total phosphorus.

All rations must be fed *ad libitum* and mash or meal forms are preferable.

Performance Characteristics

By 6–7 weeks quail will weigh some 227g (½ lb) but the range is likely to be 150 to 230g. Food consumption in this period is 454g(1lb), thus a feed conversion of 2.25 is obtained which is similar to broiler chicken performance. Mature quail consume some 3½–4 oz feed per week when laying; i.e. 15g/day.

Processing and Marketing

Killing is best carried out either by bleeding or dislocation of the head from the neck. Plucking by the scalded water method or by dry plucking is most usual with the water temperature being 148^OF (64^OC) for 15 seconds in the former method. Wax plucking can be used and gives a good finish to the carcass. For best flavour, evisceration should be left for 36–48 hours following de-feathering as this allows a game flavour to develop.

Packs of finished quail should be as evenly matched as possible. Pairs, fours or multiple of two quail are most usual. Quail is a luxury meat and benefits from being promoted in the gourmet market and through the promotion of its delicate flavour.

Diseases of Quail

Ulcerative enteritis (quail disease) is most common in Bob White quail but is also found in Japanese birds. It is a serious a condition characterised by haemorrhage, necrosis and ulceration of the intestinal wall. It is most frequently found where quail are overcrowded and mis-managed; in other words it is a disease of bad management. Symptoms are listlessness and a hunched back posture. Mortality is highest in the 6th to 8th week but can continue for several months.

The antibiotic, *streptomycin,* is used in treatment. Quail suffer from other common diseases of fowl including Newcastle Disease, Coccidiosis, Blackhead, Infectious Bronchitis and Salmonellosis.

Guinea Fowl

Guinea fowl originated from the West-coast country of Africa, Guinea. Although there are many varieties it is the speckled or pearl type which is most popular in the U.K. Very hardy birds, guinea fowl can be reared under a wide variedy of management conditions. They are perfectly happy to roost in the trees and forage for their food by day. Yet on the other hand they can be successfully reared under intensive conditions provided the correct food, water and floor space is allocated and light intensity kept low to discourage flightiness.

Breeding Stock

The normal time for Guineas to start laying is April, ceasing in the Autumn.

During this time they produce 80 to 200 eggs each depending on strain. The use of artificial light to give an artificial spring light pattern will cause them to lay in seasons other than the normal one.

Adult birds weigh some 1 to 1½ kg each at 15—16 weeks of age but careful selection of heavier type strains can lead to an 40—50% increase over this figure.

Sexing guinea fowl is virtually impossible. The males have better developed head parts, but considerable experience is needed to differentiate. Body weight is similar for both sexes. The hens have a different call note from the male but this is hardly likely to help in sexing in the commercial sense.

Guinea fowl eggs weigh about 35 to 50g, the best commercial weight being 45—50g. For best hatching results eggs of all similar size should be incubated. This also ensures an even chick size. Breeding stock can be managed either in floor pens or cages. The floor space allowance is 10 birds per square metre whilst caged birds can be controlled in about half this space. The recommended light pattern for breeders is 8 hours from day old, progressing to 17 hours by 12—13 weeks. Day length is then cut back to 8 hours over the 13 weeks to 28 weeks period. As laying commences around this age the day length is again increased up to 17 hours at the rate of ½ hour per week. Although natural mating can be used for floor managed breeders artificial insemination is used for caged stock. Insemination rate for females is about 1000/hour whilst semen collection is made twice weekly.

Incubation period is 28 days.

Rearing

Chick brooders may be used for rearing guinea fowl with the advantage that 25% more chicks can be housed. Young guinea fowl are called *keets*.

The best brooding temperature is 95°F (35°C) gradually reduced to 70°F (21°C) by 6 weeks of age when weaning can take place.

Free Range Rearing

Confining young guinea fowl is difficult but wing clipping or pinioning will prevent them flying over 2 metre high netting. Housing is

Intensive Guinea Fowl production (Photo: *Poultry World*, reproduced with kind permission).

virtually unnecessary under this system provided tree cover is available. However in fox prone areas housing is necessary. Feeding under this system can be a basic grain ration after the brooding period when a good quality chick starter is fed.

Intensive Rearing

Keets reared for the table are sold for market at 11 to 16 weeks of age depending on local market demand. Under deep litter conditions of housing each keet is allowed $0.065m^2$ of floor area for $11-12$ week rearing and 0.093 for 16 week marketing.

The birds are kept in the house from day old. Floor brooding systems used for poultry are eminently suitable and the principles followed are very similar. Up to 4 weeks of age guinea fowl are reasonably quiet but after this age they frighten very easily and care should be taken in carrying out routine chores for fear of causing a pile up in the corners with smothering and losses from suffocation.

Feeding

In commercial guinea fowl raising, economics play the most important role. Therefore whilst an ordinary chick starter ration will grow guinea fowl it does not fatten them in the time permitted; i.e. $11-12$ weeks. To do this a broiler starter crumb diet to 5 weeks, followed by a finisher pellet diet to killing age, is necessary. Range reared birds may be fed chick and poultry grower rations respectively, the latter ration being diluted by small quantities of grain in birds reared to 16 weeks of age. Apart from the foods mentioned no other feedstuff is used. Very small quantities of granite grit may be provided to help digestion.

Breeding birds should be fed a turkey breeder ration $3-4$ weeks before egg laying is anticipated at 28 weeks of age.

This ration is fed throughout the breeding period.

Marketing

Many commercial flocks of guinea fowl are raised under contract to a packing station. They should never be raised for market purposes unless an outlet has previously been arranged. To do so only upsets the supply-demand position and causes uncertainty in prices.

Some packers sell the birds in the feather after a period of 'hanging' to increase the meat's game flavour. Birds are also sold for oven ready production; in which case they are wet plucked in the normal way.

DISEASES

Guinea fowl suffer from several diseases common to chicken. Among them is coccidiosis, the intestinal parasitic infection which causes high morbidity and mortality in affected flocks. For this reason a coccidiostat should be used continuously in the fattening feeds for intensively reared stock. The coccidiostat sodium monensin is *toxic* to guinea fowl when used at the level recommended for chickens.

Chapter 12

MARKETING POULTRY

There is no room in the poultry industry for unplanned and uncoordinated production. It is vitally important for producers to make advance arrangements for the disposal of their birds at the appropriate time, for, the more control the industry can gain over the marketing and distribution of its produce, the better it becomes for each individual producer.

PROCESSING

Wet plucking is today recognised as the better method for defeathering poultry. On the smaller poultry units dry plucking by machine may still be popular owing to the lower capital outlay involved and reduced depreciation of plant machinery (Fig. 12.1). Today, in the larger packing stations, the operation of processing is fully mechanised, for several thousand birds may have to be processed every hour.

After killing, bleeding and hanging until the birds' reflex actions cease they are immersed in fast moving or agitated water. The period of agitation varies, depending on the size of the bird and the temperature of the water. For broilers and capons the period is usually 30–40 seconds in 'semi-scald' water at a temperature of $128-130^{o}F$. For the 'hot-scald' method, which removes feathers, epidermal skin and stubs the temperature is $10^{o}F$ higher. De-feathering or plucking is effected by means of rubber fingers followed by 'stubbing' or finishing in which the birds are made ready for the evisceration process and deep freezing. With the 'hot-scald' method 'finishing' is usually unnecessary, the birds being transferred immediately to deep freeze cabinets following eviscer-ation (removal of internal organs).

The object of storage is to preserve the product in a clean and suitable condition for as long as possible. Before storage is undertaken the temperature of the carcasses must be reduced to less than 50°F as rapidly as possible. Unless this is done 'greening', which detracts from the appearance of the carcass, may occur. Cooling may be carried out by blowing cold air on to the carcasses or may be effected by immers-ing them in ice slush.

The freezing process is carried out in one of the following ways:

(a) Blast freezing (- 15 to - 35°F)
(b) Liquid freezing (0°F)
(c) Contact freezing.

PRESENTATION

The way in which poultry meat is presented is vitally important if sales are to increase. The prepacked carcass whether frozen or fresh, must be attractive to the buyer; it must have 'eye-appeal'.

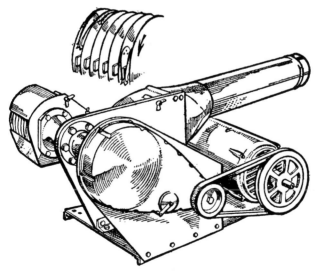

Fig. 12.1 A dry plucker which has whirring plates behind a guard.
The bird is held close to them and the feathers are rapidly stripped off

Pull away the crop of the bird. Twist it between the fingers and then pull well in order to loosen the internal organs

Cut across the vent and tailpiece to help in the insertion of fingers for the loosening and withdrawal of the intestines

Loosen the intestines. Hook fingers round gizzard, liver and heart and gently withdraw the whole of the internal organs

Needle and twine are passed through the thighs from one side to the other and the string tied tightly close to the hocks

Here is the bird finished ready for the oven except for the stuffing. This is filled in at the crop end and the small piece of loose skin remaining there serves as a flap. It can be stitched in position

Basic stages in preparing a bird for the oven. Variations are possible; including the use of rubber bands for tying and vacuum packing in plastic bags which include a brand name, thus improving the market image of the product.

Future expansion in the poultry meat industry will depend upon high standards of marketing, particularly in the presentation of the product.

It is unfortunately true that many producers take no steps to find the most satisfactory market for their products. Nor do they study the requirements of prospective buyers sufficiently. Only in the broiler industry, and to a lesser extent the turkey industry, have marketing problems been given urgent consideration. Many small producers involved in turkey, capon and table duckling production continue to rely upon the local market or dealer for the disposal of their birds. At other times, notably on festive occasions, they forward consignments of ungraded birds to the central markets. All who contemplate producing table poultry on a commercial scale should make a close study of marketing requirements and problems.

The various sales outlets for poultry products are:

(a) Consumers at the farm gate or deliveries
(b) Retail outlets
(c) Hotels and other catering establishments
(d) Packing stations and travelling dealers
(e) Large central markets

Sale to consumers

This involves dressing the products for the table. It is possible to offer a 'farm-to-table' service with this outlet provided quality standards are maintained. This outlet is very satisfactory for the smaller producer.

Sale to retailers and caterers

This usually involves direct sales to butchers, fishmongers and some multiple stores. The poultry should be graded and supplied in non-returnable cartons. To ensure continuity of supply some form of contract is usually necessary to safeguard both producer and retailer. Price should be based on birds of comparable quality in a nearby market. The catering trade usually favour the bigger birds so that a greater number of portions can be carved and served.

Sale to dealers and packers

Selling through dealers is both hazardous from a marketing angle and dangerous from the disease point of view. The enticement appears to be a cash payment. Dealers operate most successfully in times of shortage or when supply and demand is fairly even. They invariably select only the 'cream' and leave the remainder to be dumped on the nearest packing station. This is not the recommended practice and most unfair to other producers.

The poultry packing station fulfils a good need, and where the station is efficient at both processing and marketing the lure of the dealer is usually resisted. Packing stations usually operate on a service charge which includes collection from the farm, processing and marketing costs. Most noteworthy stations have developed their own brand name and it is in the producer's interest to see that only first-quality birds are marketed.

Sale to central markets

The central markets are commonly regarded as a dumping ground for surplus goods which cannot be sold elsewhere. Inferior birds are not required by the markets; they will fetch rock bottom prices.

DISTRIBUTION COSTS

In the course of wholesale poultry meat distribution several stages are encountered. At each stage a service is performed which has to be paid for and added to the retail price. The producer usually sells to the packer, who collects the birds, plucks, cools, grades, packs and freezes them and maybe stores the carcasses, before delivery to the wholesaler. The total charge for this depends upon distance and numbers collected. Many packers work on this basis, returning to the producer the price the birds realise less a service charge.

Chapter 13

EGG QUALITY

INTRODUCTION

Quality is of interest to all egg producers, for it not only affects their pockets but also the market in general.

Eggs are classified according to their weight. The following shows the U.K. Egg Authority and EEC weight classification:

Very Large	73g. and over
Large	63 - 73g.
Medium	53 - 63g.
Small	53g. and under.

This is a simplified approach from the earlier grading.

Assessment of egg quality can be made in three ways:

(a) Candling to assess internal quality
(b) Internal inspection for odours
(c) Cooking and taste tests

The Process of candling

In order to see the internal contents of the egg, the intact egg is passed over a bright light which shines through the shell and contents, outlining any faults which may be present. With modern machinery the egg is rotated over the light. This movement spins the yolk and allows the operator to see any foreign body attached to it.

Candling

The person candling eggs is called a candler and assesses their quality from:

(a) Appearance – the egg must be normal in shape, clean and flawless.

(b) Shell – an absence of hair cracks and misshapen and weak shells; it must be of even texture, strong and normal in shape (thin, brittle or porous shells are downgraded as loss of moisture and entry of bacteria lead to spoilage).

(c) Air space size – an increase in normal air space size is an indication of staleness; it is associated with the frequency of egg collection and storage conditions.

(d) Yolk – the size and position of yolk is important; a faulty yolk is sided, sunken or stuck to the shell membranes.

(e) Egg white – the white should be translucent, free from meat and blood spots, cloudiness or mouldiness.

The process of candling is a good assessment of the quality of an egg, but its accuracy is dependent on the candler's ability and judgement.

FACTORS AFFECTING EGG QUALITY

Shape

The shape is of considerable marketing significance. Egg shape is highly heritable and is influenced by selection of breeding stock. Abnormal egg shapes are often associated with diseases such as Chronic Respiratory Disease, infectious bronchitis, fowl pest and other respiratory infections.

Shell

The shell strength is most important as it governs resistance to breakage. Egg shell quality is inherited. It is also influenced by feeding, of which calcium is the influential mineral. The diet should contain sufficient calcium to meet the body function and egg production requirement. The minerals phosphorus and manganese and vitamin D3 are also involved in the calcium metabolism of the hen.

Temperature also affects shell thickness, high temperatures decreasing it owing to a decrease in blood calcium level.

Large automatic egg packing station showing the grading machinery

Shell thickness tends to decrease as the birds age, and shells may be thinner than normal as pullets come into lay. Repeated frights may give rise to thin or soft shells.

Yolk

Although yolk colour has no bearing whatsoever on the nutritive value of the egg there is a strong demand by the public for the 'golden-coloured' yolk. Colour is influenced by certain feed ingredients and is formed by the carotenoid pigments, principally xanthophylls, which occurs in grass meal, maize meal and maize products. Synthetic xanthophylls may be used, e.g., apo-carotenal or other colouring agents. Feeding cottonseed meal which contains 'gossypol' results in olive-coloured yolks and a pink tinge in the albumen and must be avoided. Mottled yolks may be caused by physical movement during the formulation of the egg in the bird. Diseases such as infectious bronchitis may also cause mottled yolks and storing eggs at too high a temperature.

BLOOD AND MEAT SPOTS

Blood and meat spots are caused by small haemorrhages at the time an ovum is released from the ovary into the oviduct. Blood spots may also result from the rupture of small blood vessels in the oviduct. Hereditary causes are not out of the question. Age appears to influence the incidence of blood spots with older birds producing more meat and blood spots.

ALBUMEN

The nature of the albumen is of considerable importance to the housewife's quality assessment of an egg. Eggs with a high proportion of firm albumen are the consumer's preference; watery whites are disliked by the housewife. Watery whites usually develop on storage at high temperatures. The height of the albumen is measured in Haugh units, the higher the figure the better or higher the albumen. High albumen height is a heritable factor. Respiratory diseases detrimentally influence albumen quality.

THE POULTRYMAN'S ROLE IN MAINTAINING EGG QUALITY

Cleanliness

The farmer has control over factors of cleanliness. To obtain a high percentage of naturally clean eggs the following points should receive attention:

(a) Adequate nesting space — one nest box should be allowed for every four birds; lack of nest space leads to dirty eggs and a higher percentage of cracked eggs.

(b) Cleanliness of nests — nesting material needs changing every week.

(c) Suitable nesting material — straw and wood shavings are ideal; rollaway nests need no litter.

(d) Frequency of collection — at least four collections per day should be made.

(e) Broodiness — this encourages cracked and dirty eggs.

(f) Battery egg production — the wire floors of battery cages should be kept clean, the egg cradle brushed down daily and droppings adhering to the wire scraped off.

(g) Storage — eggs should be stored at a temperature of 45–55°F and a relative humidity of 75%.

DRY CLEANING EGGS

Eggs which are dirty or stained should, for preference, be dry cleaned. For small numbers hand buffing with steel wool or sandpaper is satisfactory; for large-scale dry cleaning motor-driven abrasive wheels are preferable. There are many makes on the market which are capable of cleaning 1,000 eggs or so per hour.

WASHING EGGS

If egg washing is carried out two vital factors concerned with egg quality are important. First, the temperature of the water, which should be about $80-90^{\circ}F$; higher temperatures will cause albumen coagulation. Secondly, a sanitiser should be used only with very dirty eggs, and even then not for more than 5 minutes. Water should be frequently changed. Dirty water defeats the object of cleanliness and disease control.

Chapter 14

COST OF PRODUCTION

Keeping accurate records is an essential part of modern poultry keeping. The economic position in all sections of the poultry industry is such that profit margins are low, further emphasising the need for maintaining records of production. This chapter sets out to provide the interested reader with facts and figures typical of those found on the efficient poultry farm.

REARING REPLACEMENT LAYING PULLETS

The cost of rearing can have a marked effect on the cost of egg production. Indeed, the cost involved less the carcass value of the culled hen can amount to one sixth of the total egg production costs. Modern pullet rearing is a highly specialised business. It can be broken down into two sections, brooding and rearing. Production costs vary considerably, depending on the type of pullet, method of housing and system of feeding. Food accounts for approximately 50% of total costs, while chick and food costs combined amount to 70–80% of total costs. Heating and lighting are comparatively cheap but nevertheless very important cost items.

EGG PRODUCTION

Feeding accounts for about 70% of the costs involved in egg production; thus, any saving in food wastage will materially affect the food cost. Egg production costs are measured in many ways, the most important

Footnote: **The prices of foodstuff and labour will inevitably change so costs in this chapter are given to demonstrate the principles, which can be adapted.**

of which is the cost to produce 12 eggs. This, of course, can only be found when all the costs are accurately recorded. It is, however, possible to arrive at an approximate total cost by finding the food cost per 12 eggs and assuming that this represents 70% of the total costs.

The following illustration shows the breakdown in costing for a pullet recorded over a 52 week period.

	pence (p)
Point of lay pullets	140
Food	470
Electricity	10
Labour	30
Depreciation	30
Mortality	12
	692

In this illustration, involving a medium-weight type of pullet, the food consumption should be about 41 kg with an average daily food consumption of 120 g per bird. Mortality should be no more than ½% per month under good management conditions. The food cost per 12 eggs is found by dividing the number of dozen eggs laid into the food cost.

Records of grading should be kept and the various costs of producing each grade calculated at the end of the year. In so doing, the producer will know at what time of the year his pullets should be purchased in order to obtain maximum profitability.

BROILER PRODUCTION

Broiler production is a highly specialised industry involving very low production costs and profit margins. For this reason large numbers of birds are kept under highly intensive management conditions. The producer should not attempt broiler production unless he has a guaranteed outlet for his produce.

Facts and figures are very numerous, making comparisons between different strains and feeding techniques relatively easy. Typical results are shown by the following figures, which refer to broilers, as hatched at 52 days and 48 days of age.

Age sold	48 days	52 days
Area per bird	0.037 m^2	0.040 m^2
Average liveweight (kg)	1.8	2.0
Mortality	3.2%	3.5%
Food consumption per bird (kg)	3.60	4.2
Feed conversion ratio	2.0:1	2.10:1

Production costs other than food are very low. This is shown in the following illustration.

	pence/bird
Chick	11.00
Litter	0.70
Heat and light	2.75
Medication	0.07
Vaccination	0.20
Labour and management	2.15
Insurance	0.50
Clean out	0.31
Miscellaneous	0.30
	17.98

On the larger broiler units there is little variation between the figures given. Small producers will, however, find their costs proportionately higher mainly because their purchasing and bargaining powers are restricted.

Feed Costs

Feeding costs are constantly increasing and can be calculated by multiplying the cost per kilogram of food by the amount consumed per bird. Thus, if food cost per kg is 12p then food cost per bird is 12p x 4.2 kg (50.4p) for a 52 day broiler. The food cost per kg of meat is found by dividing the food cost per bird by the average weight. Food and chick costs amount to 88–90% of the total costs involved in producing broilers.

CAPON PRODUCTION

Modern capon production usually involves two weights of bird. First, with the use of broiler cockerels and high-efficiency feeds, capons can be produced weighing 3.2 kg in 12 weeks of age. Secondly, capons weighing 4.5 kg are produced in 18–20 weeks. The first type also involves intensive housing, while the second type may be produced under less intensive conditions, depending on the time of year. Costings are extremely variable because of the variation in bird, food and housing used. The following figures must therefore be used as a rough guide only.

	12–week capons *pence (p)*	*18–week capons* *pence (p)*
Chick	10.00	10.00
Food	77.00	140.00
Heat and light	3.00	4.00
Medication	0.25	0.50
Miscellaneous	2.00	3.00
Labour and management	3.00	6.00
	95.25	163.50
Cost per kg meat	30p	40p

TURKEY PRODUCTION

Basically, two age groups or types of turkeys are now produced:

(a) Teenage birds marketed from 10 to 16 weeks.

(b) Christmas table birds marketed at 18 to 26 weeks of age.

Teenage turkeys marketed at 16 weeks of age should weigh approximately 7.0 kg liveweight (mixed sexes). The weight will naturally vary depending on feeding, housing and strain of turkey used. Food conversion rate should be in the order of 2.5:1.

Because the male poults have a higher protein requirement than the hens the sexes should be separated and the birds fed and reared separately. To encourage a good finish on birds marketed under 16 weeks of age the birds may be caponised once, 4 weeks before killing.

The illustration below gives typical production figures for the two types of turkeys:

	15 week (pence)	20 week (pence)
Poult	66.0	66.0
Feed	250.0	340.0
Labour	12.0	15.0
Electricity/heat	12.0	13.0
Litter	5.0	6.0
Miscellaneous	5.0	5.0
	350.0	445.0
Cost per kg meat	50.0	50.0

TABLE DUCKLING PRODUCTION

Table ducklings, as pointed out in Chapter 9, have a limited market and only those producers with a guaranteed outlet should attempt this type of production. Strains of duckling vary in their ability to put on weight and convert food into flesh. Producers contemplating this production should therefore make sure that they purchase the best type of duckling for the market's requirements.

The following production figures will serve as a guide to producing table duckling between 7 and 8 weeks of age weighing 3 kg liveweight.

Duckling	40.0
Food	100.0
Labour	3.0
Electrictiy/heat	2.0
Litter	0.5
Miscellaneous	5.0
	150.0
Cost per kg meat	50p

Broiler Breeders

Parent stock for producing commercial broilers are stocked at around 0.02 m² to 0.023 m² per hen housed. The full life cycle is about 63 weeks, inclusive of a 4 week clean out period. The total period is made up of a 0–24 week rearing period and 25–59 week breeding period.

Cost of Production

The cost of producing hatching eggs is calculated from the following information.

	pence per bird
Feed (0 to 59 weeks)	580.00
Labour and Management	110.00
Heat/light	25.00
Litter	8.00
Medication	1.00
Vaccines	2.20
Insurance	6.50
Miscellaneous	5.00
	737.70
D.O. Chick cost	74.00
Total cash costs	811.70
Depreciation cost	90.00
Total costs/hen housed	901.70

Cost per 12 hatching eggs = 82p
(132 hatching eggs per hen) = 902p
132 hatching eggs/hen is 'break even' point.

Chapter 15

HYGIENE AND SANITATION

INTRODUCTION

In no aspect of poultry husbandry does the intelligent use of scientific knowledge show greater dividends than in the control of disease. Scientific investigation has made tremendous strides in recent years, mainly as a result of the advent of specific drugs or what is known as chemotherapy. This has been offset to some extent by the increase in the occurrence of certain diseases as a result of intensivism.

Intensive forms of management, such as in battery houses and build-up or deep litter, may have much to recommend them as labour-saving systems but, nevertheless, they increase the risk of the spread of infectious diseases and parasites as a result of the denser populations, especially where communal and automatic feeding and watering devices are in use. Respiratory diseases, in particular, spread more rapidly under fully intensive conditions, where there is little dilution of droplet or airborne infection by outdoor air.

Disease control, therefore, cannot be achieved without a sound knowledge of husbandry and hygiene. The correct spacing of birds, adequate ventilation, the prompt recognition and isolation of ailing birds, proper management of litter and general sanitation of housing and equipment are just as important, if not more so, than knowledge of modern drugs.

It must be remembered, too, that subnormal health resulting from chronic infections, parasitism and faulty nutrition, interfering with full production and efficiency, can take an even greater toll of profits than the more spectacular and acute killing diseases.

Certain economic surveys have shown that approximately 10% of birds die in their first laying year, while a large proportion succumbs when the fowl nears or reaches maturity. In the second instance the farmer has already incurred the cost of purchasing or breeding the chick and of feeding and rearing it to the time when egg production should start, and these costs are not offset by income from egg production or poultry flesh.

Until recently the advice for most outbreaks of infectious disease has been to destroy infected or ailing birds, depopulate and disinfect. This advice, still sound in many instances, was based on the fact that the low value of the individual fowl was such that the health of the flock as a whole was of more importance than that of any of its component units, and treatment of the individual was uneconomic and dangerous.

The introduction of vaccines and specific drugs that can be used in the food or water has made the prevention and treatment of a number of diseases a safe and practical proposition. It has meant that the accurate diagnosis of the nature of the disease is even more important than hitherto, for these new drugs, highly powerful in their effect, are equally highly selective in their action against different disease agents and are not 'cure alls'.

A drug highly efficient against one disease may be equally valueless against another. Its faulty use, therefore, resulting from wrong diagnosis, may be not only wasteful but also dangerous in that the disease may go unchecked.

RECOGNITION OF SYMPTOMS

One cannot labour the point too strongly that the first cardinal law in hygiene or disease control is the rapid recognition of a sick bird and the establishment of an accurate diagnosis. Fortunately, it is usually possible in poultry keeping to sacrifice a typically affected bird for a post mortem examination. Although the poultry keeper should acquaint himself with the lesions or changes which accompany the commoner diseases so that temporary measures can be put in force, he should as far as practicable always obtain a qualified diagnosis, for many diseases can be differentiated only by laboratory techniques.

Hygiene is an important part of battery routine. This plant is being pressure sprayed with steam before the new batch of pullets goes in

The live bird is, of course, the greatest danger in transmitting disease and, in addition to the foregoing precautions, it must be remembered that adult fowls often harbour parasites without themselves showing symptoms and that they can transmit such infection to the more susceptible chicks. Chicks should be reared, therefore, as far as is practicable, in isolation from adult stock, and separate rearing houses and equipment should be kept for this purpose. Traffic of attendants between the old and young stock should be reduced to the minimum. These precautions appear to be particularly valuable in the control of viral infections.

CARRIERS OF DISEASES

In a similar way certain species of poultry can carry diseases which affect others. The caecal worm of the fowl, for example, is responsible for spreading the parasite that causes blackhead in turkeys and for this reason it is undesirable to attempt to rear turkeys either in contact with fowls or on ground where poultry has recently been kept.

The carcasses of fowls dying from an infective disease are also common source of infection. They should be removed immediately from the house and never left lying about to be attacked by vermin which could in turn transmit infection to the rest of the flock. For preference, such carcasses should be incinerated. Lime pits are sometimes used but are not so efficient. Carcasses should never be left lying on the manure heap.

Neighbours and other farmers should be discouraged from bringing sick birds or the carcasses of dead birds on to one's own premises for examination, and visiting other premises in order to advise on outbreaks of disease should also be avoided. That is the province of the veterinary surgeon, and in any case there is the danger that you may transmit infection back to your own stock.

Rats and mice, apart from being carriers of certain diseases such as salmonellae infections, can do a lot of damage and cause serious waste. Food stores should, as far as possible, be vermin-proof.

Clean egg production is essential in the control of eggborne disease, since bacteria in dirt and droppings can spread through the pores and infect the embryo, which in turn will affect other chicks during hatching – this quite apart from the lower returns from packing stations for dirty eggs or the labour involved in cleaning them.

This is an important point, for no egg, by whatever method it is cleaned, is ever as clean bacteriologically as an egg which is laid clean. Many methods of cleaning eggs increase the risk of spoilage during storage, since the washing of an egg in any fluid of a lower temperature than that of the egg causes bacteria to be drawn through the pores of the shell. Wiping with a damp cloth always hastens penetration of bacteria, while the cloth itself becomes contaminated and spreads infection to other eggs. A danger with washing machines is that the brushes may become contaminated and transmit infection to subsequent eggs.

Dry cleaning, although laborious, is probably preferable to wet cleaning or, alternatively, dirty eggs can be dipped into germicidal solutions containing a detergent. It is essential to remember, however, that the solution so used must be at a temperature higher than that of the egg i.e. $26.5^{\circ}C - 32^{\circ}C$. Clean egg production can be improved by collecting eggs at more frequent intervals and seeing that nest box litter is renewed frequently.

SPREAD OF DISEASES

Most infectious diseases are spread by food or water which becomes contaminated with infected droppings or other discharges. With the exception of the built-up or deep litter system, therefore, droppings should always be treated as being potentially infected.

Some parasites – coccidia, for example – are not infective immediately they are passed out by the host but must spend a period outside the bird's body before they can infect a new host or re-infect the same bird. With coccidia this period is, in the most favourable conditions for the parasite, a minimum of 48 hours.

Warmth and humidity favour the multiplication of such parasites, and damp areas in litter should be avoided either by moving water and feeding troughs frequently or by using one of the sanitary types of feed and water troughs which prevent birds having access to their droppings.

Grass in runs should be cut short so that parasites and other infective agents get a maximum exposure to sun and light, rapid drying being one of the surest methods of destroying infection. Coccidia, on the other hand, can resist lower temperatures, remaining in a quiescent state, but will again multiply and become infective when climatic conditions are favourable. This is the reason why outbreaks recur on farms from season to season. Worm eggs are also highly resistant and remain infective for long periods in the litter or soil.

Disinfection of the soil is difficult and uncertain, and the best method of dealing with infected ground is to plough or dig over the land, lime heavily and reseed, leaving it vacant for the longest possible period.

External parasites such as lice and mites appear to multiply most quickly in a dark, humid atmosphere. Houses, therefore, should be well ventilated and well lit, and they should be so constructed that no part can be the permanent harbour for dust and dirt. Perches, nest boxes and droppings boards are the most favoured sites for external parasites such as mites and lice and these should be movable so that the fittings can be dismantled for regular cleaning.

In the same way, walls and floors should, as far as is practicable, be smooth and free from cracks so that they can be easily washed. Treating the walls, floor and ceiling with creosote, lime washing or paraffin emulsions incorporating chemicals is an effective method of controlling external parasites.

DISINFECTION

A thorough disinfection of all poultry houses should be carried out at least twice a year and always before new stock is moved into the house. It must be remembered that most disinfectants lose their efficiency in the presence of organic matter such as grease and manure, while bacteria and viruses are so minute that they can be protected in the smallest piece of dried excreta. The first stage, therefore, in disinfecting a house is to sweep up all litter and refuse, scrape the floor, walls and ceiling and remove the accumulated sweepings to the manure pit. When there has been infection present in the house it is advisable to first spray with a strong disinfectant then, having collected the litter and sweepings, to burn them.

The empty house should then be scrubbed with hot water containing 4% washing soda or a detergent. A high-pressure pump is the most suitable for this purpose since it will penetrate normally inaccessible places. After drying, the house should be sprayed with an approved disinfectant in a strength recommended by the makers.

Built-up litter, of course, is a different proposition, and it depends entirely on the use to which this is being put as to the steps that should be adopted. With adult birds the litter can be used for a number of seasons provided that disease has not been present and the litter has been well managed. It is inadvisable, however, to attempt to rear young stock on built-up litter that previously has maintained older birds.

With chick rearing for broilers it is still debatable whether the litter should be removed and renewed between each batch or heaped in piles approximately 1 m high and 2 m at the base for a period of about a week before being respread. Heaping of built-up litter does produce temperatures capable of destroying parasites such as coccidial oocysts and worm eggs. Even when litter has been heaped, however, it is advisable to spread fresh litter under the hovers. Regular turning of the litter helps to turn fresh droppings to the base, where they will be exposed to quantities of ammonia lethal to coccidia.

Generally speaking, so far as the control of parasites is concerned the deeper the litter the better, right from the start. Shallow litter increases the risk of setting up outbreaks of disease.

Chapter 16

DISEASES AND PARASITES [*]

VIRAL

NEWCASTLE DISEASE (Fowl Pest)

Cause

N. D. is caused by a myxovirus.

Transmission

The virus is excreted by infected birds from the respiratory tract and in the droppings. Can be spread from farm to farm by dust particles containing virus or by mechanical transfer on equipment, vehicles, personnel or by wild birds. The incubation period varies from 2 to 14 days.

Species Affected

Mainly chickens and turkeys.

Symptoms

Young birds — Signs of disease are difficulty in breathing. The chick opens the beak and gasps for breath. This can cause a rattling noise and coughing. The condition spreads rapidly through a flock and leads to nervous symptoms which may be seen as trembling or as partial or complete paralysis. Twisting of the neck is said to be characteristic of this infection.

Adult birds — The respiratory symptoms are not very clear. The condition spreads rapidly through an unvaccinated susceptible flock. Food intake is reduced and in laying birds egg production can fall to near zero. Malformed eggs will be seen. Egg production gradually returns

* See Editorial Note at end of chapter.

over 6–8 weeks but rarely returns to normal.

Mortality

Young birds –A virulent field infection can lead to 90% mortality in chicks.

Adult birds – Mortality will vary with the virulence of the infection.

Diagnosis

It is necessary to look at the total picture to arrive at a diagnosis. Clinical information and post-mortem findings must be combined with laboratory aids.

Treatment and Control

There is no substitute for vaccination. There are no known drugs which will overcome the N.D. virus. If a flock is infected then antibiotics may be given in order to reduce the danger from secondary invaders (bacteria).

INFECTIOUS BRONCHITIS (I.B.)

Cause

I.B. is caused by a virus.

Transmission

The virus excreted from infected birds may pass to other sites by the airborne route. The virus may also be transported on clothing or machinery. The incubation period varies from 18 to 48 hours.

Species Affected

Chickens are susceptible to this disease.

Symptoms

Young birds – The main signs of disease are nasal discharge, gasping and coughing. Sneezing may also be heard in the early stages. Female chicks under 2 weeks of age, when infected with virulent I.B. virus, may suffer permanent damage to the oviduct and become false layers in later life.

Adult birds – Gasping and coughing will be seen. In laying birds a marked drop in egg production occurs and there will be a high percentage of misshapen and rough egg shells. In addition when the egg is broken the albumen is watery. Recovery can take several weeks and production rarely reaches the level anticipated. Virulent I.B. can be the main trigger for chronic respiratory disease.

Mortality

In chicks below 6 weeks of age mortality can reach 25%, but in older birds mortality is negligible.

Diagnosis

There are three main factors to be considered in order to arrive at a diagnosis:

    (a) The clinical picture including post-mortem findings in the flock.

    (b) Isolation of the virus in the laboratory.

    (c) A rising antibody titre when the serum is tested against a known strain of bronchitis virus.

Treatment and Control

Prevention of the disease by vaccination with attenuated live virus is the best approach. There is no specific therapy for the virus infection although secondary infection can be controlled with antibiotics.

INFECTIOUS LARYNGOTRACHEITIS (I.L.T.)
(Chicken influenza)

Cause

I.L.T. is caused by a virus belonging to the avian herpes group.

Transmission

Natural infection occurs by way of the respiratory tract and consequently airborne virus is the common means of spread. Mechanical spread on equipment and by wild birds is of relatively minor importance. The incubation period varies from 4–12 days.

Species Affected

The chicken is the main host, but the virus can also effect pheasants.

Symptoms

The disease spreads rapidly through a flock and causes difficulty in breathing. The head may be extended and the beak open to improve the air intake and there may be some coughing which can result in the birds bringing up blood-stained mucous from the trachea. Death frequently occurs due to blockage of the trachea with blood and exudate. In laying flocks a drop in egg production will be seen which can vary from 10–60%. Normal production is resumed after about 4 weeks.

Mortality

The number of deaths in a given outbreak will depend on the health status and management of the flock and also on the virulence of the field strain. Under some conditions mortality is as low as 5% whilst in others it can be as high as 70%. The average is about 12%.

Diagnosis

When the report of the disease has indicated a rapidly spreading respiratory infection with some of the birds coughing up blood and mucus it is already a good indication of I.L.T. Post-mortem examination usually reveals haemorrhagic tracheitis.

Treatment and Control

There are no satisfactory treatments and control should be aimed at ensuring a high standard of management with vaccination of breeding stock. Vaccination can be achieved using a live eye-drop vaccine but other strains suitable for drinking water administration are being developed.

FOWL POX (Avian Pox)

Cause

Fowl Pox is caused by a virus.

Transmission

The existence of 'carrier' birds which are capable of acting as the source of infection seems to be the most likely means of transmission. Mosquitoes and other flying insects have been shown to spread the virus. The incubation period varies from 4 to 20 days.

Species Affected

Chickens, turkeys, pheasants, pigeons, etc.

Symptoms

The lesions normally occur on the head and in the mouth of the bird. It is also possible to find lesions on the legs and mucous membranes such as the cloaca. Lesions on the head and comb are usually of a wart-like nature. Mouth lesions are diphtheritic and have the appearance of a cheesy membrane. The general symptoms will depend on the area of the body affected and birds will appear dull. When the mouth is affected they may have difficulty in breathing and eating.

Mortality

When the virus attacks the mucous membranes of the nasal cavities mortality can reach as high as 40–50%. The virus causes a marked increase in the output of mucus. The less serious cutaneous form of the disease can cause a reduced egg production.

Diagnosis

Wart-like lesions of the head particularly of the comb and around the eyes. In some birds a diphtheritic membrane may be seen in the mouth. Laboratory confirmation should be obtained.

Treatment and Control

It is difficult to treat affected birds. It may be possible to remove the diphtheritic membrane from the mouth and larynx thus improving respiration. The disease is best prevented using live pox vaccine administered either by wing web or feather follicle application.

AVIAN ENCEPHALOMYELITIS (A.E.)
(Epidemic Tremor)

Cause

A.E. is caused by a virus.

Transmission

Much work has been done to determine the way in which infection is transmitted. The virus can be egg-borne and this reduces hatchability. Chicks that do hatch wll show symptoms of the disease and pass on the infection in the incubator to newly hatched clean chicks. Young chicks can also be infected on the farm. The incubation period varies from 5 to 14 days dependent on the route of infection.

Species Affected

The main host is the chicken although the pheasant can also be infected.

Symptoms

The disease is mainly seen in chicks one and three weeks of age. Affected chicks crowd on their haunches and move in a spasmodic manner. Some of the chicks may fall on their sides, paralysed. A slight trembling of the head and neck may be seen. Hence the name Epidemic Tremor. In laying and breeding flocks A.E. may cause a marked drop in egg production (up to 60%). Production is resumed after about 3 weeks but will never reach its normal level and eggs from breeding flocks will show low hatchability.

Mortality

Apart from the reduced hatchability of infected eggs, chicks infected naturally suffer high mortality (75%).

Diagnosis

When the clinical picture is studied in detail together with post-mortem examination of typically affected birds the veterinarian can usually decide on the cause of the trouble. Histological examination of the central nervous system will show typical changes. A laboratory serum-virus neutralisation test has been developed which must be carried out in A.E. antibody free eggs.

Treatment and Control

No satisfactory treatment is known. When a problem has been established in chicks from a particular breeding flock it is advisable to stop hatching eggs from this flock for a time. Vaccination of future laying and breeding stock is possible using live virus vaccines.

LYMPHOID LEUCOSIS
(Big Liver Disease, Avian Leucosis, Visceral Lymphomatosis)

Cause

The virus causing this disease is a tumour-producing myxovirus.

Transmission

The virus has been shown to be transmitted through the fertile egg to the developing chick. (Vertical transmission). Transmission can occur between young birds through saliva and droppings. (Horizontal transmission).

Species Affected

The main species affected is the chicken.

Symptoms

The disease normally affects laying birds at the point of lay or very shortly afterwards, although infection will have occurred either through the egg or in the first weeks of life. Affected birds will appear listless and weak. The abdomen is distended and on palpation it will be possible to feel an enlarged liver.

Mortality

This can range from 5% to 40% and will depend on many factors not least of which is the strain or type of bird. Some strains of bird are more resistant to the disease than others.

A victim of fowl pox, a contagious disease, revealed in one form by small watery blisters on neck, wattles, and around the corners of the beak, which tend to dry into brown crusts, sometimes running together to form large cauliflower-like growths

These birds show two common symptoms of fowl paralysis — paralysis of the legs and dropped wings

Diagnosis

Post-mortem findings are characteristic. The liver is so enlarged that it can appear to fill the abdomen. It is granular and grey in colour. The ovaries are frequently affected and appear to be studded with grey tumours. The kidneys may also be infiltrated by tumours. The heart and other internal organs and membranes may be affected with tumours. During the last decade several laboratory tests have been developed to demonstrate the presence or absence of the disease.

Treatment and Control

There is no satisfactory treatment and there is no vaccine available. Under very strict conditions of management it is possible to keep birds free of Lymphoid Leucosis. The two areas where there are commercial possibilities to control the disease are genetic selection and vaccine development.

MAREK'S DISEASE (M.D.)

(Fowl Paralysis, Neural Lymphomatosis, Ocular Lymphomatosis)

Cause

This disease is characterised by tumours and paralysis caused by a Herpes-type virus, now known as Marek's Disease Virus (M.D.V.)

Transmission

There is no evidence that the virus can pass from the parent through the developing egg, to the offspring (Vertical transmission). The disease spreads readily from bird to bird and is highly contagious. Cells or feather follicles appear to be the main way in which spread occurs.

Species Affected

The domestic fowl.

Symptoms

The classic symptom is paralysis with the bird unable to stand or walk. The skin may be thickened by tumours. Tumours of a yellow/ white colour can also be found in the internal organs. Sometimes the iris is affected and appears white, hence the name ocular lymphomatosis.

Mortality

This is the most serious disease facing the poultry industry in nearly all countries of the world. Mortality can range from 5 to 60% with the average about 15%. Losses are nearly always seen between the 10th and 20th week of the bird's life, but cannot occur after laying has started.

Diagnosis

Paralysed birds. Whenever there is high mortality in birds between 10 and 20 weeks of age Marek's disease should be suspected.

Post-Mortem Findings

Skin and feather follicles may be thickened and leathery. One or both of the sciatic nerves may be enlarged. Tumours may be found in the ovaries and other internal organs.

Treatment and Control

Being a virus disease there is no satisfactory system of treatment. It is possible to rear birds in special houses and to keep them free of the disease. Vaccination can be used, using the Herpes Virus vaccine of the turkey origin (HVT).

GUMBORO DISEASE (G.D.)
(Infectious Bursal Agent)

Cause

This disease is caused by a virus and is sometimes referred to as the Infectious Bursal Agent (I.B.A.)

Transmission

There is not a great deal known of the precise method of transmission, but the disease is recognised in many countries of the world and in some countries is of increasing importance. Likely means of spread is by the birds, by equipment or people associated with birds. An infected house can remain infected when empty for several months.

Species Affected

The domestic fowl.

Symptoms

Birds become inactive and have a reduced appetite. Many show trembling and quivering with the feathers in a ruffled condition. Affected birds pass watery faeces. Mortality occurs quickly.

Mortality

Up to 20% of a flock may be affected and show symptoms which disappear after 5–10 days. Mortality is dependent on the age of the birds and can vary from 1–15%. At 3–5 weeks of age mortality averages 5%. In older birds mortality tends to be lower.

Diagnosis

The clinical picture is important in providing an indication of the disease affecting the birds. Confirmation is obtained by post-mortem examination. The most striking findings are haemorrhages in the musculature of the legs, swelling of the kidneys and a marked enlargement of the bursa of Fabricius.

Treatment and Control

Antibiotics and chemotherapeutics have no effect on the condition. Vaccines have been made using attenuated field strains of virus. Good management in the form of strict hygienic measures must be maintained.

BACTERIAL

AVIAN TUBERCULOSIS (T.B.)

Cause

Caused by a bacterium: Mycobacterium tuberculosis avium.

Transmission

The disease is contagious and is passed from bird to bird by contact with the droppings of infected birds.

Species Affected

The domestic fowl.

Symptoms

The disease normally affects older birds. There may be progressive weight loss and the birds become unthrifty. The muscles of the breast are reduced in size exposing the sternum or breast bone. Appetite usually remains normal until the terminal stage of the disease. There may be some lameness and swelling of the joints. The comb and wattles appear pale in colour.

Mortality

The disease is not responsible for many deaths. The number of birds in a flock which are affected can vary from 5 to 95%.

Diagnosis

Post-mortem examination reveals lesions which appear as yellow/white caseous nodules in the liver, spleen and intestines. Smears made from the lesions can be examined and will show the bacilli.

Treatment and Control

There is no satisfactory treatment. The disease may be eliminated by good management. Affected flocks should be destroyed and buildings and equipment sterilised thoroughly before restocking.

INFECTIOUS SYNOVITIS
(Infectious Arthritis)

Cause
The bacterium Mycoplasma synoviae.

Transmission
The way in which birds become infected is not known but there is evidence to suggest transmission through the egg. Once a flock is infected the disease spreads very slowly. Stress conditions such as other diseases, vaccination and sudden changes in climatic conditions aggravate the disease.

Species Affected
Domestic fowl and the turkey.

Symptoms
Birds affected with Infectious Synovitis are lame and the comb is pale. Growth is retarded. Swellings may be seen around the joints of the feet and legs (bumble feet). Birds become listless, emaciated and dehydrated. Droppings may have a green discoloration. Breast blisters are common.

Mortality
The numbers of birds which die as a result of this infection is normally very low (less than 1%). However, the number of birds infected can be high (up to 75%).

Diagnosis
The presence of lame birds in a flock with swellings of the feet and joints. The comb may be pale in colour and the birds will be listless. Droppings may be watery and green in colour. When the swellings in the feet are incised a creamy exudate may be seen from which the M. synoviae organisms can be isolated and identified.

Treatment and Control
Treatment of affected birds is not practical. Birds showing symptoms should be removed from the flock. Litter should be dry and non abrasive. The remaining birds should receive oxy- or chlortetracycline at a level of 200 ppm for at least 2 weeks. Furazolidone is also beneficial. Any birds subsequently showing symptoms must be removed from the flock.

STAPHYLOCOCCAL INFECTION
(Staphylococcosis, Staphylococcal arthritis)

Cause
The disease is caused by the bacterium Micrococcus aureus.

Transmission
The bacteria causing the disease may gain entry to the body through wounds or through the alimentary tract.

Species Affected
The disease has been reported in chickens, turkeys, ducks, geese and game birds.

Symptoms
In the acute form the birds die without symptoms. If the birds survive the acute phase, then the bacteria can settle in joints making them enlarged and painful. The birds become very lame and unable to walk and may die of starvation due to their inability to reach the food troughs.

Mortality
The disease is more widespread than is recognised and mortality can be as high as 30%.

Diagnosis
By the isolation of Micrococcus aureus from the blood or joints of affected birds, together with the clinical picture.

Treatment and Control
Broad-spectrum antibiotics may have a beneficial effect on the disease. However it is far more important to pay attention to things such as sharp surfaces, particularly where birds are being kept on wire. Housing and utensils should be examined for defects which might cause mechanical damage to the birds.

INFECTIOUS CORYZA
(Haemophilis gallinarum infection, Roup)

Cause
The bacterium causing this disease is Haemophilis gallinarum.

Transmission
Disease spreads from bird to bird by contact and airborne infected

dust or droplets and the drinking water. Spread by equipment and personnel has also been reported. The incubation period varies from 1 to 3 days.

Species Affected
   Only chickens.

Symptoms
   The main signs of the disease are inflammation of eyes and nose with foul-smelling discharges, conjunctivitis, sneezing and facial swellings. Feed and water intake are reduced, leading to loss of weight. Egg production in laying birds declines.

Mortality
   Mortality will vary but is generally low.

Diagnosis.
   Field inspection produces similar symptoms to Chronic Respiratory Disease, a diagnosis is difficult to establish. Diagnosis may be obtained by the isolation of the organism from the sinus or air sac exudate from affected birds. This procedure is carried out in the laboratory.

Treatment and Control
   Prevention is the best approach. Recovered birds must be removed from the flock. They remain carriers of the infection. Antibiotics and chemotherapeutics reduce the severity of the disease. Attention must be paid to ventilation and to hygienic measures.

FOWL CHOLERA
(Avian Pasteurellosis)

Cause
   It is caused by the bacterium Pasteurella multocida.

Transmission
   Infection occurs by way of the respiratory and alimentary tracts. Can be spread by wild birds and the body excretions of diseased birds which contaminate water, soil and feed. Mechanical transfer on equipment and vehicles is also possible. Stress conditions increase the susceptibility to infection. The incubation period varies from 4 to 9 days.

Species Affected
   Practically all species of poultry.

Symptoms

Peracute stage — Many dead birds. Cyanosis and swelling of the comb and wattles are typical.

Acute and Chronic Stage — Swelling of joints and legs causing lameness. Birds are listless and refuse to eat or drink. The disease causes difficulty in breathing and a thick nasal discharge. A greenish-yellow diarrhoea may be observed.

Mortality

In the peracute stage mortality may run as high as 90%, in the acute and chronic stage losses up to 60% are seen.

Diagnosis

The isolation of Pasteurella multocida from the heartblood or from liver tissue of affected birds, together with the clinical picture.

Treatment

Antibiotics and chemotherapeutics have been reported to have a beneficial effect. It is far more important that strict hygienic measures and the control of visitors and equipment be introduced. Buildings and equipment should be disinfected thoroughly before restocking. Vaccines have been developed with variable results.

PULLORUM DISEASE
(Bacillary White Diarrhoea, B.W.D.)

Cause

The bacterium Salmonella pullorum.

Transmission

The bacteria can be transmitted to the fertile egg by an infected parent and hence infect the day old chick. These chicks can then infect other chicks in the incubator, in chick boxes or under the brooder. Breeding stock that have survived infection go on to complete the cycle. In addition to this cycle of transmission the disease can be carried by infected equipment and staff.

Species Affected

A wide range of birds including:— Chicks, pheasants, ducks, geese and guinea fowl.

Symptoms

Chicks: Chicks hatched from infected eggs may die shortly after hatching in the incubator. Chicks that survive are listless and some may be moribund. They huddle together and frequently exhibit a white diarrhoea. Many of the birds will die of acute septicaemia.

Adults: When a parent bird is suffering from the chronic form of infection there are no outward signs of the disease. Acute infection can occur in adult birds who show loss of appetite, weakness and a greenish-brown diarrhoea.

Mortality

Chicks: The number of deaths will depend on a number of factors including whether the birds are infected in the incubator or in the brooder. Under some conditions mortality can reach 50%.

Adults: Mortality is not normally very high.

Diagnosis

It is essential to demonstrate the organism by bacteriological examination of the tissues of dead chicks. In older birds internal organs such as the heart, gizzard, lungs and intestines may show grey nodular lesions. The ovaries of affected hens may be enlarged, irregular in shape and brown-green in colour.

Agglutination tests carried out on blood taken from birds in a flock suspected of infection can be of great help in establishing a diagnosis.

Treatment and Control

Treatment with sulphonamides and drugs like furazolidone have been shown to reduce losses but are far from satisfactory as a complete cure. Control is the best approach to the problem. This is done by only breeding from flocks that have been cleared of infection. If the eggs are kept free of infection and the incubators, brooders and other equipment are kept clean the birds will remain free of infection.

PROTOZOAN AND PARASITIC

COCCIDIOSIS

Cause

Coccidiosis may be caused by many different species of Coccidia. Each species gives rise to a particular pathological picture with characteristic symptoms.

The coccidia affecting chickens of the genus Eimeria have a life cycle which is dependent on the bird. The cycle is divided into an asexual phase and a sexual phase.

The asexual phase:

The birds are infected by viable oöcysts. In these oöcysts the sporozoites mature and on release invade the intestinal wall of the bird. There they develop and become mature trophozoites (or schizonts). Each mature schizont contains numerous merozoites which escape and in their turn invade a new intestinal cell. They develop to new schizonts containing again merozoites (2nd generation).

Sexual phase:

The second generation of merozoites enter new cells and develop themselves to male (microgametocytes) and female (macrogametocytes). The microgametocytes penetrate the macrogametes (matured macrogametocytes), the resulting zygotes forms a wall about themselves and the oöcysts thus developed escape in the faeces of the birds ready to infect other chickens. When ingested by a new host the life cycle is complete.

Coccidiosis can be divided into 2 groups:

1.   The caecum is involved (Caecal coccidiosis).

Mainly caused by E. tenella in chickens up to 12 weeks. Mortality may be as high as 50%. Birds are listless, have bloody droppings, a pale comb and a lack of appetite. Laboratory examination will show haemorrhages in the caecal wall.

2.   The small intestine in involved (Intestinal coccidiosis).

Caused by E. acervulina, E. brunetti, E. maxima. E. necatrix.

E. acervulina

Affects birds of any age. E. acervulina is not normally pathogenic, but in some cases considerable mortality may be seen. Birds infected show loss of weight, combs may be shrivelled and a drop or even cessation of egg production in layers may be seen.

E. brunetti

May affect birds of any age. E. brunetti is definitely pathogenic, in severe infections mortality can be high. Birds infected show emaciation and diarrhoea.

E. maxima

May affect birds of any age. E. maxima is less pathogenic, mortality is generally low. Diarrhoea, loss of weight, and a drop in egg production of layers, will be seen; bloody droppings are commonly seen.

E. necatrix

Mainly in chickens up to 4 months of age. E. necatrix is very pathogenic. At the acute stage mortaiity may be high in the first week after infection. In the chronic stage blood may be seen in the droppings, the birds are listless, and lose weight. In layers a drop in egg production will be observed.

There are other species of Eimeria occurring in chickens, e.g. E. mitis, E. praecox, E. Hagani, which are of less importance.

Treatment and Control

There is no disease group in poultry where both control and treatment are employed more.

The well established principles of good management and husbandry are of basic importance.

- cleanliness of personnel and equipment
- good ventilation
- avoid overcrowding
- separate young and old animals
- disinfection before restocking

It is common practice to include low levels of a coccidiostat in the feed. These chemicals keep in check the development of the parasites so that a pathological situation does not develop. It should however be taken into account that coccidia may develop a resistance to all chemicals and for this reason it may be necessary to change from one chemical to another periodically. Treatment of infected flocks may be carried out by the administration of coccidiostats at a higher therapeutic level. There are certain products (coccidiostats) available which are specifically designed for treatment and which are not satisfactory for prevention. Whenever administering these products particular attention should be paid to the dosage recommendations of the manufacturer.

BLACKHEAD
(Enterohepatitis, Histomoniasis)

Cause

Blackhead is caused by the protozoan Histomanas meleagridis.

Transmission

Though direct transmission by contaminated water, feed or droppings is possible, the main source of infection is the embryonated egg of the caecal worm (Heterakis gallinarum) ingested by the birds.

Species Affected

Turkeys, chickens and game.

Symptoms

Turkeys: Usually young birds up to 3 months of age. Also older birds. Characteristic symptoms are listlessness, weakness, ruffled feathers, emaciation and a sulpher-coloured diarrhoea. Sometimes black colouration of head parts will be seen – hence its name!

Chickens: The symptoms are not as marked as in turkeys. Birds show a period of unthriftiness; yellow droppings may be seen. Chickens are most susceptible at 4–5 weeks of age.

Mortality

Turkeys: Losses may be as high as 100% and averages about 50% unless kept under control. Mortality will be heaviest in poults during the first 3 months of life.

Chickens: Losses will vary with the virulence of the infection but will generally not be higher than 10%.

Diagnosis

The liver may be swollen, often with typical lesions; petechial haemorrhages will be found in the heart muscle. Microscopic examination of lesions of the caecal wall. The causal organism can only be seen in very fresh material examined within one hour of death.

Treatment and Control

Turkeys: Avoid contact of young chickens with poults by keeping them separate.

Chickens and Turkeys: Strict hygienic measures should be taken on farms where regular outbreaks are seen. Medicated feed or water should be given. The drinking water is regularly refreshed. Anthelmintics should be used to eliminate the worms which act as the vector.

INTESTINAL PARASITES

LARGE ROUND WORM
(Ascaridia galli)

Symptoms

This is a large parasite up to 8 cm in length which causes loss of condition, reduced egg production in layers and when infection is heavy, death.

Retarded growth, listlessness and diarrhoea are the symptoms. At necropsy the large round worms can be seen in the middle part of the small intestine.

Treatment and Control

It is very important to see birds are not exposed to the intermediate hosts (e.g. snails, earthworms). Good management and adequate hygienic measures are essential.

In flocks where the parasites have been found specific anthelmintics should be administered.

HAIRWORM
(Capillaria)

Symptoms

This small parasite affects the upper part of the alimentary tract causing diarrhoea, weakness, reduced egg production and anaemia.

Affected birds are listless, weak and lose weight.

Treatment and Control

It is very important to see that birds are not exposed to the intermediate hosts (e.g. snails, earthworms). Good management and adequate hygienic measures are essential.

In flocks where the parasites have been found specific anthelmintics should be administered.

DEFICIENCY DISEASES

CRAZY CHICK DISEASE
(Vitamin E deficiency)

Symptoms

Vitamin E deficiency results in encephalomalacia (brain disorder) in chicks. The gizzard musculature is also involved, so that the birds affected are eventually unable to feed or walk. The hocks are enlarged and the legs bowed. At post mortem the cerebellum (area of brain) is enlarged and shows small haemorrhages.

Treatment and Control

Adequate levels of Vitamin E should be included in the ration.

Should treatment be necessary Vitamin E should be administered by mouth or by the water.

Adequate dietary protection should be given by using a suitable antioxidant in the feed in order to spare the vitamin's antioxidant role.

RICKETS
(Vitamin D3 deficiency)

Symptoms

Vitamin D3 deficiency results in retarded growth in chickens, leg weakness and soft beaks and bones. The ribs are curved inwards. Awkwardness of gait will be observed.

In affected layers there is a marked increase in the number of soft-shelled eggs, and a drop in egg production may be seen.

Treatment and Control

Prevention of this disease can be assured by the inclusion of adequate vitamin D3 in the ration. Affected birds should be treated with supplementary vitamin D3.

FUNGAL

ASPERGILLOSIS
(Fungal Pneumonia, Pulmonary mycosis)

Cause

Caused by the fungus Aspergillus fumigatus.

Transmission

By contact and inhalation of spores from Aspergillus fumigatus which are found in contaminated litter. The disease is also reported as a hatchery born disease.

Species Affected

Domestic fowl, especially young birds.

Symptoms

Loss of appetite, increased thirst, gasping and accelerated breathing. Eye infections may be present.

Mortality

Generally from 10% up to 50%, especially in young chickens and turkey poults.

Diagnosis

The presence of Aspergillus fumigatus demonstrated microscopically or sometimes even by the naked eye in the air passages of the lungs, in the airsacs or in lesions of the abdominal cavity.

Treatment and Control

By eliminating the affected birds. Good management in the form of strict hygienic measures on buildings, feed, litter and equipment. Special disinfection methods are recommended. Treatment with the antibiotic nystatin has proved effective in some flocks.

VACCINES AND VACCINATION POLICIES

Important points to note

* Vaccines should be stored in the dark at approx. +4°C/39°F (domestic refrigerator).

* Each vial should be used immediately after opening.

* Birds to be vaccinated should be in good health. Sick or weak birds do not develop sufficient immunity following vaccination.

* The vaccination schemes are intended only as a guide; local conditions must be allowed for and the advice of a veterinary surgeon should be followed.

* Wash and disinfect hands after vaccinating. Any surplus vaccine should be destroyed either by burning or boiling.

Vaccine against avian encephalomyelitis (A.E.) in chickens

Use:

For the vaccination of future layers and breeding stock.

Vaccination Programme

AGE	VACCINE	ADMINISTRATION
3–4 months	A.E. Vaccine	Oral instillation/drinking water
Moulting period	A.E. Vaccine	Oral instillation/drinking water

At least 14 days interval should be allowed between two consecutive vaccinations with different vaccines.

N.B.

On no account should A.E. vaccine be used for vaccination of chicken younger than 2 months and birds in lay.

Vaccination of layers and breeding stock must be carried out at least 1 month before the beginning of the laying period, thus preventing the risk of A.E. in chichens hatched from eggs of these birds .

I.B. VACCINE

Against Infectious Bronchitis

Indication

1. For re-vaccination of future layers and breeding stock.

Normal Vaccination

AGE	VACCINE	ADMINISTRATION
3 weeks	I.B. H-120 or	
	B.P. H-120	Drinking water/installation
16 weeks	I.B. H-52	
Moulting period	I.B. H-52	

Emergency Vaccination

AGE	VACCINE	ADMINISTRATION
1–5 days	I.B. H-120 or	
	B.P. H-120	
3 weeks	I.B. H-120 or	Drinking water/installation
	B.P. H-120	
16 weeks	I.B. H-52	
Moulting period	I.B. H-52	

At least 14 days interval should be allowed between two consecutive vaccinations with different vaccines.

After an I.B. H-120 vaccination of birds of 3 weeks or older, there should be an interval of no less than 10 weeks and no more than 15 weeks before revaccinating with I.B. H-52.

2. BROILERS

Normal Vaccination

AGE	VACCINE	ADMINISTRATION
2–3 weeks	I.B. H-120	drinking water/instillation

Emergency vaccination

AGE	VACCINE	ADMINISTRATION
1–5 days	I.B. H-120	instillation/spray method
3 weeks	I.B. H-120	drinking water/instillation

3. LAYING BIRDS

When it is necessary to vaccinate birds in the laying period, I.B.H-120 should be used. Revaccination after each 3 months with the same vaccine should be carried out. At least 14 days interval should be allowed between two consecutive vaccinations with different vaccines. After primary vaccination with I.B. H-120 at 1–5 days there should be an interval of 2–3 weeks before revaccinating with the same vaccine.

Administration

Through the drinking water, by the spray method or by intranasal/intraocular instillation.

Immunity and vaccination reaction

Duration and intensity of the vaccination reaction and the establishment of a solid immunity are dependent on the possible presence of maternal antibodies and in general on the health and condition of the birds. Hygiene and management are also important in the post-vaccination period.

The reaction after primary vaccination is mild. Slight respiratory disturbances may occur 4–7 days post-vaccination. These symptoms will usually disappear within two weeks. Revaccination will normally cause no visible reaction.

Immunity will set in one week post-vaccination and be optimal a few weeks later.

Protection obtained from a vaccination, carried out at 1–5 days or 3 weeks following the directives, will be maintained for 3 or 13 weeks respectively. An emergency vaccination of non-immunized hens in lay might provoke a fall in egg-production, which will usually recover.

MAREK VACCINE

Administration

The dose for each bird is a single intramuscular injection of 0.1 ml.

No repeat injections are necessary.

Recommended age for injection is one day but Marek Vaccine can be effectively administered to birds up to three weeks old.

Once diluted, the vaccine should be used within six hours. Any unused vaccine should be discarded after this time.

Storage

The freeze-dried vaccine should be stored in a refrigerator at $+2^{\circ}C$ to $+10^{\circ}C$.

The diluting fluid may be stored at room temperature.

VACCINE

Against Newcastle Disease (N.D.) in chickens

Strain LaSota

Indication

For re-vaccination of future layers and breeding stock.

Normal Vaccination

AGE	VACCINE	ADMINISTRATION
2 weeks	Hitchner B1	
or 3 weeks	or H-120	
8 weeks	LaSota	drinking water/instillation
22 weeks	LaSota	drinking water/instillation
Moulting period	LaSota	drinking water/instillation

Emergency Vaccination

AGE	VACCINE	ADMINISTRATION
1−5 days	Hitchner B1	(see specific literature)
4 weeks	Hitchner B1	(see specific literature)
or		
1−5 days	H-120	(see specific literature)
3 weeks	H-120	(see specific literature)
10 weeks	LaSota	drinking water/instillation
24−26 weeks	LaSota	drinking water/instillation
Moulting period	LaSota	drinking water/instillation

At least 14 days interval should be allowed between two consecutive vaccinations with different vaccines. After vaccination with Hitchner B1 or H-120 at 2−4 weeks there should be an interval of 6−7 weeks before revaccinating with LaSota. The interval between two consecutive LaSota vaccinations should be no less than 3½ months and no more than 5 months.

N.B.

Primary vaccination of susceptible hens in lay should never be carried out with LaSota. Use Hitchner B1. Revaccination of hens in early egg-production may be carried out.

Administration

Through the drinking water or by intranasal-intracular instillation.

Immunity and vaccination-reaction

Duration and intensity of the vaccination-reaction and the establishment of a solid immunity are dependent on the immunity obtained from the primary vaccination and in general on the health and condition of the birds. Hygiene and management are also important in the post-vaccination period.

Revaccination will normally cause no visible reaction when the directives are followed. Protection obtained from the first LaSota vaccination carried out at 8−10 weeks, will be maintained for at least 4 months. The second LaSota vaccination will give an immunity lasting for one laying period of normal duration.

VACCINATION METHODS

I. *Drinking water administration*
This method saves time and labour.

* Not to be used for chicks younger than 5 days.
* Vaccination reaction may be seen after 5 days.

Reconstitution of Vaccine
a. Water is withheld from the birds before the vaccine administration.
b. Prepare sufficient water for the number of birds to be vaccinated.
c. Open up the correct number of vials under the surface of the water.
d. Mix thoroughly with clean spoon ensuring that all vials used are emptied.
e. Offer to birds immediately.

For more detailed information concerning the administration of specific vaccines consult the directions enclosed in each carton.

Administration
The water should be withheld for 2–3 hours before vaccination. Where possible, divide the administration of the medicated water into two parts giving the second part 30 minutes after the first, so ensuring that all birds receive the correct dose during the time the vaccine medicated water is available.

Points to Observe
1. Use fresh rain water OR clean, cold tap water in which chlorine or metals can neither be tasted nor smelt. (To prolong the life of the virus liquid skim milk may be used. This may be added to the water at the rate of 600 ml (approx. 1 pint) per 10 litres of water (approx. 2 gallons). After mixing well the solution should be allowed to stand for 15–30 minutes before mixing the vaccine into the water. Only SKIM MILK should be used as the fat in whole milk causes problems of blocking in automatic drinking systems).
2. Ensure that all the vaccine medicated water is consumed within 2 hours.

3. It is important to provide a sufficient number of water containers to ensure adequate drinking space for each bird. All containers should be clean and free from any traces of soap, detergent or disinfectant.

4. Climatic conditions may render it advisable to vary the period for which water is withheld prior to vaccination. i.e. the higher the temperature the shorter the period.

5. When the number of birds is between standard dosages, the next higher dosage should be chosen.

6. Wash and disinfect hands after vaccination. Contaminated equipment should be sterilised by burning, boiling or by immersing for a few hours in a strong disinfectant solution.

Calculation of dosage rate:
Simple Drinking Troughs and Fountains

Number of doses	For 2–4 week old birds Quantity of water	For birds over 4 weeks of age Quantity of water
1000	10–20 litres (approx. 2–4 gallons)	20–40 litres (approx. 4–8 gallons)

II. Intranasal – Intraocular Instillation
   * For emergency vaccination of chicks under 3 weeks of age.

Reconstitution of Vaccine
The vaccine should be dissolved in normal saline solution.

III. Spray Method of Vaccination (for day old chicks)
The main advantages are: –
1. Early protection,
2. Simple application,
3. Reduced stress during subsequent vaccinations,
4. Protection independent of maternal antibody level,
5. Low vaccination costs.

The Spray Method should only be used with the following vaccines:
Vaccine HITCHNER B1
Vaccine H-120
Vaccine HITCHNER B1/H-120

The spray vaccination is carried out at the hatchery or at the poultry farm IMMEDIATELY after the arrival of the chicks in boxes. Spray method can be used for some secondary vaccinations but veterinary advice is necessary.

Detailed technique for the application of the Spray vaccination

I.   The poultry house should be cleaned and disinfected before the chicks arrive.

II. The hatchery should be contacted to ascertain exactly at what time the chicks will arrive. Everything needed for the spray vaccinations can be prepared.

III. The chicks have to be vaccinated immediately after arrival.

IV. If the spray vaccination is carried out in the chick boxes, the boxes should be placed in a row side by side in the poultry house. In diffuse light the boxes are opened and the chicks vaccinated.

DO NOT PUT THE CHICKS UNDER BROODERS OR NEAR OTHER HEATERS DURING OR SHORTLY AFTER THE SPRAY VACCINATION, (a distance should be kept between the birds and the brooders of 1.5 m).

V.   Check the quality of the day-old chicks. They must be healthy.

VI. Preparation of the vaccine solution and filling of the spray apparatus:

a.   ONLY PURE OR DISTILLED WATER AT A TEMPERATURE OF 25°C should be used.

b.   1000 doses are dissolved in 0.5–1.5 litres water by opening the vial under water. The number of birds to be vaccinated with this volume will depend on local conditions.

* EDITORIAL NOTE: Methods and drugs are constantly changing and, therefore, readers in doubt should consult a veterinary surgeon.

Footnote: The poultry farmer must be aware of the regulations regarding registration of flocks and the compulsory vaccination against *Salmonella enteritidis*.   Contact the Department of Agriculture for details.

Index

Numbers in italic refer to the page numbers of illustrations

Index

Index